T0069449

Landscapes and Geomorphology: A Very Short Introduction

VERY SHORT INTRODUCTIONS are for anyone wanting a stimulating and accessible way into a new subject. They are written by experts, and have been translated into more than 45 different languages.

The series began in 1995, and now covers a wide variety of topics in every discipline. The VSI library now contains over 500 volumes—a Very Short Introduction to everything from Psychology and Philosophy of Science to American History and Relativity—and continues to grow in every subject area.

Titles in the series include the following:

Andrew Goudie and Heather Viles

LANDSCAPES AND GEOMORPHOLOGY

A Very Short Introduction

OXFORD

UNIVERSITY PRESS

OXFORD

UNIVERSITY PRESS

Great Clarendon Street, Oxford OX2 6DP

Oxford University Press is a department of the University of Oxford.
It furthers the University's objective of excellence in research, scholarship,
and education by publishing worldwide in

Oxford New York

Auckland Cape Town Dar es Salaam Hong Kong Karachi
Kuala Lumpur Madrid Melbourne Mexico City Nairobi
New Delhi Shanghai Taipei Toronto

With offices in

Argentina Austria Brazil Chile Czech Republic France Greece
Guatemala Hungary Italy Japan Poland Portugal Singapore
South Korea Switzerland Thailand Turkey Ukraine Vietnam

Oxford is a registered trade mark of Oxford University Press
in the UK and in certain other countries

Published in the United States
by Oxford University Press Inc., New York

© Andrew Goudie and Heather Viles 2010

The moral rights of the author have been asserted
Database right Oxford University Press (maker)

First published 2010

All rights reserved. No part of this publication may be reproduced,
stored in a retrieval system, or transmitted, in any form or by any means,
without the prior permission in writing of Oxford University Press,
or as expressly permitted by law, or under terms agreed with the appropriate
reprographics rights organization. Enquiries concerning reproduction
outside the scope of the above should be sent to the Rights Department,
Oxford University Press, at the address above

You must not circulate this book in any other binding or cover
and you must impose the same condition on any acquirer

British Library Cataloguing in Publication Data

Data available

Library of Congress Cataloging in Publication Data

Data available

Typeset by SPI Publisher Services, Pondicherry, India

Printed and bound by
CPI Group (UK) Ltd, Croydon, CR0 4YY

ISBN 978–0–19–956557–3

9 10

Contents

List of illustrations

Chapter 1
The changing landscape

The landscapes that we see around us are complex and multilayered, and have often developed over almost unimaginably long timescales. Understanding how these landscapes evolve and change, and why they are important both globally and locally, is what this book is all about. We are going to take you on a journey through time and space to find out more about the landscape and the scientists (known as geomorphologists) who investigate it. But, before we start on this journey, we need to think more carefully about what landscapes really are. For many of us, the best view we get of a landscape is as we break through the clouds coming in to land after a flight. On the approach to Heathrow Airport near London, for example, we see rivers, fields, roads, villages, and towns draped over a topography of low, rolling hills. Underneath these features are the rocks that underpin this landscape – which have been formed, altered, and contorted over millions of years of geological history. On exposure at the Earth's surface, these rocks have become shaped by water, wind, and ice, producing the topography (or relief) we see today, as well as the soils which blanket the landscape. Both relief and soils are slowly changing – adapting to new conditions as climatic and other environmental factors change. Vegetation and animal life have spread across these surfaces, leaving their own, continuing, impact on soils, relief, and climate. Finally, humans have left, and continue to leave, their sometimes indelible imprint on the landscape – through buildings, transport routes, field systems, quarries, and other constructive and destructive activities. Often, what is most visible about a landscape are the upper layers of vegetation and human society, but the underlying surfaces are not blank, smooth canvases – rather, they are rough, diverse, and dynamic surfaces which play an active role in creating the look and function of a landscape. So, landscape is what is termed a 'palimpsest' – a series of complex and overlying layers. In the landscape, these layers also interact, unlike those in the original meaning of a palimpsest, which is overprinting of writing on an ancient manuscript or parchment.

There are many dramatic and spectacular examples of landscape palimpsests which illustrate the overlying layers and their complex evolution. New Zealand, a mecca for geomorphologists, possesses a whole range of spectacular landforms which include alpine mountains, enormous fault lines, and coastal fjords. The fundamental reason for the development of such grand scenery is the presence of colliding lithospheric plates. New Zealand is, in fact, a fragment of the old supercontinent of Gondwana which formed around 650 million years ago, and started to break up around 130 million years ago. New Zealand probably detached from Gondwana around 80 million years ago, and during the last 65 million years or so has become the site of the boundary between the Pacific and Indian/Australian plates. Today, a great fault line – the Alpine Fault – runs south-westwards through South Island and marks this boundary. There is much compression across this fault, and this creates the Southern Alps, which at Mount Cook reach an altitude of 3,764 metres. These mountains have been rising very rapidly, possibly by as much as 20 millimetres per year, and have formed over the last 5 million years. The Southern Alps are also shaped by the processes of glaciation, weathering, and erosion that are manifestations of the climate. Their crests trap moist air that comes from the Tasman Sea, so that in exposed areas there are massive annual precipitation amounts, up to 12,000 millimetres. This means that exceptionally large glaciers have formed. In the

Landscapes and Geomorphology

2

Pleistocene (the period from about 2 million to 12,000 years ago), these glaciers greatly expanded in volume during glacial periods and carved the troughs in which the great fjords, which serrate the southern coast of New Zealand's west side, developed. The geographical isolation of New Zealand since it split from Gondwana has had great impacts on the fauna and flora, as did the arrival of European colonists in the early 19th century. Today, the New Zealand landscape reflects this complex history, spanning hundreds of millions of years and linking the forces of tectonism and climate, and the physical and the living components of the landscape (see Figure 1).

Because landscapes consist of rocks, soils, vegetation, animals, and human constructions, many different groups of scientists and academics are involved in their study and management. In many countries, geographers have claimed landscape as their key area of study – as geography has, over its long history, commonly focused on human/environment relations, or what we might see as the upper layer of the landscape palimpsest. Ecologists,

1. **Landscape around Arthur's Pass, South Island, New Zealand**

geologists, archaeologists, and historians have also laid claim to landscape as one of their major subjects of interest, as they focus on different layers or timescales of landscape change. However, our purpose with this book is to introduce you to how and why geomorphologists, as the main 'landscape scientists', study landscape.

Geomorphology may be defined as the study of the Earth's surface and the processes that shape it. Thus, by definition, it deals with the fundamental 'canvas' of landscape – that is, the topography, or relief, and the processes that create and shape it. Geomorphology is what holds landscape together. As befits such a central area of study, geomorphology is carried out by scientists trained in a range of disciplines, notably geology and geography. Geomorphology is also today a highly inter-disciplinary field, with linkages to hydrology, ecology, climatology, and human geography, for example. What geomorphologists do, and how they do it, will be a central theme of this book.

Geomorphology is a complex science, very different from the experimental sciences of physics and chemistry. Like geology and biology, geomorphology is often described as an 'historical science', as it deals with change over time. The argument goes that in experimental sciences, such as chemistry, an experiment (as long as it is run properly) will always produce the same outcome wherever or whenever it is run, in observance of basic laws. However, in historical sciences, such as geomorphology, an event may have a very different outcome depending on the conditions at that particular time and location. For example, an earthquake of the same magnitude and location will not necessarily produce the same landslide event within the same area of mountainous terrain at two different dates, because all other things are not equal – landscape has a history and that history makes each event unique. This makes geomorphology both fascinating and frustrating, as it is very hard to find general laws within landscapes and thus very difficult to explain and manage them.

The quest for general explanations of how landscapes behave is one key motivation for geomorphologists, as we shall see later on.

Whilst geomorphology is clearly complex and difficult, at heart it focuses on three fairly simple elements: that is, landforms, processes, and the development of landscapes over time. As a subject, geomorphology has sometimes prioritized one of these three over the others, but we now know that they are all crucial to explaining landscape. These are the building-blocks of geomorphological science which together create what is often seen as the 'geomorphological system', or 'Earth surface system'. In these phrases, 'system' implies an interconnected, functioning association of different elements. Systems, as a concept, are often used in the historical sciences to try and make sense of dynamic and complex subject material. They both organize and simplify the essential components of geomorphology, helping geomorphologists to see what is significant and what is not. But before we look at geomorphic systems in any more detail, let us explain what the three building-blocks of geomorphology are all about.

Some landforms, such as river valleys, mountain ranges, and beaches, are familiar. Others are more obscure features that may be given locally coined names, such as dolines, playas, tafoni, and yardangs. In essence, landforms are clearly defined topographical features. Occupying part of the Earth's surface, they have a three-dimensional shape and are usually made up of sediments and rocks, water, and organic life. If we fly over any part of the Earth's land surface (in a plane or, virtually, on Google Earth), we will see that the topography is organized into recognizable landforms. Geomorphologists, especially during the 19th and early 20th centuries, have spent much time identifying, measuring, naming, and explaining these landforms. Often, local words have been used to name landforms in the different countries in which they occur (like 'doline', which is a term for a closed depression found in limestone landscapes originating in a Serbo-Croat name for 'valley'), and much effort has been exerted trying to create internationally standardized terminologies. Philosophically aware geomorphologists (and there aren't a large number of these!) have also devoted time to considering whether landforms are true entities or 'natural kinds'. Similar discussions have also been held in biology over the concept of 'species' and whether a species is a natural kind. Landforms are, in many ways, the species of the geomorphological world. Even non-philosophically aware geomorphologists, however, realize that landforms are rather more elusive and slippery to identify and categorize than biological species. Individual landforms do not have a unique signature – there is no geomorphological equivalent of DNA.

Landforms can be big or small. Indeed, they range hugely in scale from small pits in rock surfaces less than a centimetre in diameter to whole mountain ranges of thousands of kilometres in extent. Big landforms may be made up of many small ones. A large river basin, such as the Amazon, is a landform, and one which has superimposed on it many smaller landforms (such as river meanders, sand bars, hillslopes). Geomorphologists have investigated in many different ways whether the small landforms contribute to the development of the larger ones on which they are superimposed and, in turn, whether the large landforms constrain the development of the smaller ones. As well as varying in spatial scale, landforms also vary in terms of how long they take to develop. As with spatial scale, the range of timescales over which landforms evolve varies hugely – from a few seconds to millions of years. In general, smaller landforms develop more quickly than bigger ones, but there is large variability within this general rule. One useful way of visualizing the vast variability in size and timespan of formation of landforms is to plot them on a log:log plot. Such graphs have logarithmic axes and are commonly used to graph datasets which cover very large-scale ranges and would not fit on standard axes. These diagrams are used frequently in geomorphology and other Earth and environmental sciences.

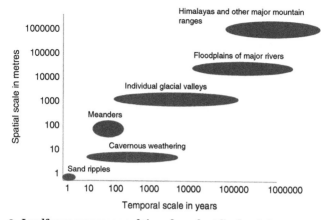

2. Landforms over space and time: from short-lived and tiny to immense and persistent

A diagram of this type with a number of landforms plotted on it is shown as Figure 2. The vast range of scales of landforms illustrated by this diagram hints at the many questions facing geomorphologists.

What sorts of questions do geomorphologists pose about landforms? Often, geomorphologists want to know what produces particular types of landform, how quickly they develop, and how persistent they are within a landscape. We also often want to know where specific landforms are found – for example, whether they develop only under specific climatic conditions and, if so, whether they can be used diagnostically as evidence to help us interpret newly discovered landscapes. We are also interested in links between landforms and life – whether particular landforms harbour specific vegetation types and, in turn, whether vegetation communities help produce some landforms. Geomorphologists are also increasingly interested in the linkages between landforms and global change – such as how sensitive some landforms may be to future climate change. Finally, geomorphologists also ask questions about the contribution of landforms to the overall topography of the Earth. We will look at many of these questions in later sections of this book.

The second major building-block of geomorphology is the study of Earth surface processes. These are the processes that sculpt the materials which make up the Earth's surface. In essence, they shape the canvas of landscape. Geomorphologists recognize two major types of Earth surface process, giving them the labels 'exogenic' and 'endogenic'. 'Exogenic' means working from outside, and is used to refer to processes fuelled ultimately by the Sun's energy and usually operating through the climate system, such as erosion by wind and waves. 'Endogenic' means working from inside, and is used by geomorphologists to refer to those processes powered by energy from inside the Earth, such as volcanic and tectonic processes. The terms used to describe these key processes are often quite imprecise and confusing (surprisingly, given their importance to the science of geomorphology), but all the processes identified contribute to change in the land surface. Many geomorphologists have regarded the development of landscape as a 'battle' over time between the two sets of processes, exogenic and endogenic. Or, to put it another way, the essence of geomorphology can be interpreted to be the long-term interplay between climate and tectonics. Let's look at this in a bit more detail.

Exogenic processes largely involve what geomorphologists call 'denudation', or the lowering of the land surface through the linked processes of weathering and erosion. For example, wind, rain, snow, ice, and gravity all contribute to eroding mountain landscapes. Gradually, and sometimes abruptly, steep mountainsides become eroded into more gentle slopes as sediment is detached, removed, and washed away through the river system, ultimately ending up in the oceans. If the Earth had no tectonic system, geomorphology would be dominated by denudation and, over millions and millions of years, the

Earth's surface would be entirely flattened as a result of the slow processes of denudation. However, the Earth does possess a tectonic system, which is responsible for the endogenic processes shaping the Earth's surface. These, in distinction to denudation processes, are predominantly constructional. Tectonic uplift produces mountain ranges, and volcanic eruptions produce new land (such as volcanic islands). As land becomes uplifted, the battle starts as slopes become steepened and denudation becomes accelerated, producing surface lowering. As material is removed by denudation, uplift may be encouraged as the crust 'bounces back'. In some parts of the world, denudation and tectonic uplift work hand in hand today, whereas in other places tectonic activity is rare today, but had an important influence in the distant past. Thus, the battle between endogenic and exogenic processes can sometimes be an episodic and one-sided affair.

Earth surface processes can be plotted on a diagram (Figure 3) to illustrate the vast range of rates at which they operate. Some processes work incredibly slowly, such as soil creep which moves sediment only a few millimetres each year. Soil creep involves the movement of individual grains of soil down a slope as a response to minor changes in microclimatic conditions. Under frosty conditions, for example, grains of soil become detached on small ice crystals developing from water in surface pores. When the frost thaws, the ice supporting the soil grain melts and the grain moves under gravity a small distance down the slope. Such conditions may only occur a few times a year, and thus the progress of an individual soil grain can be extremely slow. The area affected may also be very small, in this case just a few grains of soil. In distinction, some Earth surface processes operate remarkably quickly. The movement of material down slope under gravity (mass movement) can be very rapid, such as rockfall which can move debris many hundreds of metres within a few seconds. Such a rockfall can also cover a very large area, with several millions of tonnes of material moved in one event. Looking at Figure 3, we can see that both endogenic and exogenic processes can occur

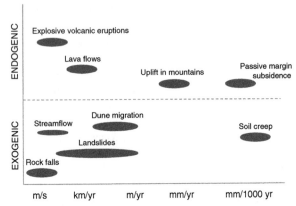

ENDOGENIC

Explosive volcanic eruptions

Lava flows

Uplift in mountains

Passive margin subsidence

EXOGENIC

Streamflow

Dune migration

Soil creep

Landslides

Rock falls

m/s   km/yr   m/yr   mm/yr   mm/1000 yr

3. Earth surface processes over time: how quickly do they operate?

slowly or quickly. They can also affect areas from centimetres to hundreds of kilometres in diameter.

What questions do geomorphologists ask about Earth surface processes? Quantifying their rate is a crucial area of research, as is identifying where and when they occur. Geomorphologists are also interested, as we have seen above, in linking processes to the development of landforms in order to understand the role they play in the landscape. We are also engaged in attempts to show how geomorphological processes interlink with ecological processes – such as for example the mutual operation of weathering and nutrient cycling. Furthermore, geomorphologists are increasingly interested in linking Earth surface processes to other processes affecting the whole Earth system, such as biogeochemical cycling and human activity. We will return to some of these key questions in later chapters.

The third building-block of geomorphology is the study of the development of landscapes over time. It is fair to say that this preoccupied geomorphologists for much of the 19th and early 20th centuries, and became neglected during the second half of the 20th century when geomorphologists (at least in the English-speaking community) became obsessed with quantifying processes over the short term and at the local scale rather than investigating the longer term. With the advent of new and improved techniques, such as radiometric dating, the study of long-term change has now been reinvigorated. So what do we mean by 'long term' in geomorphology? How long does it take for landscapes to develop? If we picture landscapes as being large-scale assemblages of landforms, then most landscapes have been produced over the Cenozoic era – the last 65 million years or so of geological time. However, change over the Cenozoic has often been influenced by events during much earlier geological periods, in which rocks have been laid down, large plate tectonic movements have occurred, and sea levels have changed dramatically. Some landscapes are extremely old, such as large parts of Australia which still bear the signs of processes and events in much earlier geological periods. Within the Cenozoic, the past 2.4 million years or so (often called the Quaternary period) have been the dominant period of change for many landscapes. More recently, human impacts on the landscape have become an important geomorphological force over the last 12 thousand or so years (often called the Holocene), with an acceleration of human activity during the past 300 years. Also over the Holocene, there has been a major period of global sea level rise associated with the melting of the ice at the end of the last glacial period. This has been a major control on the development of coastal landscapes throughout the world. The world's major deltas, like the Nile and the Mississippi, began to develop their present forms around 6,000 years ago.

Why do geomorphologists expend so much effort trying to explain histories of landscape development? One important reason is that to understand how landscapes function today, it is necessary to evaluate their history. Trying to understand the pattern and timing of catastrophic mass movements within the Himalayas, for example, requires an understanding of the forces that have shaped the landscape in the past. Glacial periods left clear imprints in the Himalayas and other mountainous areas, in the form of over-steepened valley side slopes where glaciers had eroded. These over-steepened slopes became unstable and likely to fail once the glaciers melted. Reconstructing glacial histories in time and space can help determine the likely threat of geomorphic hazards today.

The geomorphic past is also a major testing ground for our theories and ideas of how the world works which have been derived from studying the present. If, for example, geomorphologists develop a theory to explain the link between weathering and the development of cavernous weathering features (also known as tafoni), then they can test this by looking at a dated sequence of tafoni to see how they developed under different past climatic and environmental conditions. A further reason to study the broad changes in landscape in the past relates to the need to predict the future. If we can reconstruct how landscapes, climate, and tectonics have mutually adjusted over hundreds, thousands, and millions of years, we can start to understand their behaviour better and look ahead to future changes (perhaps linked to climate change). However, in order to do this, geomorphologists need to utilize all three of their basic building-blocks, and bring together a concern for landforms, processes, and landscape development over time. How can they achieve this?

In order to link these elements, it would be helpful if geomorphologists had some over-arching theory or view of how the world works. As we shall see in Chapter 2, there have been several attempts to create such a theory, and the jury is still out over whether we will ever have a satisfactory over-arching theory which suits everyone and accommodates all questions and research areas. Certainly, at the very least we need some shared goals and methods and, as we discuss in Chapter 3, geomorphology has come a long way in developing a suite of widely applicable techniques and methodologies in recent years. Putting these together has enabled geomorphologists to make much progress in interpreting past, present, and future landscapes. In Chapter 4, we discuss in more detail the impact of tectonics and climate change on landscapes, whilst Chapter 5 introduces some of the interesting work done on how the living and non-living parts of the landscape interact. As we shall see in Chapters 6 and 7, we now have a lot of information and ideas with which we can respond to environmental managers' requests for explanations of the human impacts on the landscape (both today and as a result of ongoing and future climate change). Geomorphology also has other dimensions, such as a long-term engagement with landscape and culture (introduced in Chapter 8) and its application to other landscapes, such as those on other planets and the ocean floor, as we discuss in Chapter 9.

Chapter 2
The present is the key to the past

Today's geomorphology, or landscape science, builds very clearly on the work of past scholars from a wide variety of backgrounds. It is hard to identify who the first geomorphologists were, as that 'job title' didn't exist until the late 19th century, whilst interest in the scientific study of landscape stretches back many centuries. People such as Leonardo da Vinci carried out early geomorphological studies, through their attempts to understand phenomena such as rivers. Da Vinci's sketchbooks reveal his interest in river flows and turbulence. Like da Vinci, most of the early scholars who investigated geomorphological phenomena did so as part of a much wider interest that they had in life and the Earth. Even in the 19th century, when there was a great flowering of landscape study, the scholars involved were interested in a much wider range of phenomena. Perhaps the best example of this is Alexander von Humboldt (1769–1859), who investigated a huge array of topics within the natural environment, culminating in the publication of his multi-volume work entitled *The Cosmos* which covered subjects ranging from astronomy to human societies. Others, such as Charles Darwin, gained long-term fame and recognition for their research on other parts of what we now call the Earth's system, but nevertheless also carried out important geomorphological research.

How did these early scholars come to recognize that the Earth's surface changes over long timescales? How did they come to understand the relationship between landform and process? In essence, they were early 'landscape detectives', utilizing the existing range of techniques (and helping develop some new ones) in order to do four things: describe landscapes, interpret unfamiliar ones in the light of what they already knew about the way the world works, develop theories, and then, where possible, test them by dating. It is these landscape-detection methods which have come to form the basis of modern-day geomorphological science, and it is worth examining them in a bit more detail. We are going to explore them in relation to three major 18th- and 19th-century geomorphological problems: that is, how do the different types of coral reefs develop?; are rivers powerful agents of denudation?; and what created the extraordinary landscapes of the Alps?

The first stage of being a landscape detective involves describing the features of an unfamiliar landscape. Imagine visiting for the first time somewhere like Iceland and being faced with an almost unimaginable array of odd landforms, which don't look anything like any landforms you have ever seen before. How would you go about describing them? Today we would undoubtedly take photographs, but in the past field sketches formed a key part of the landscape detective's toolkit, and even today these are an important part of most geomorphologists' fieldwork. Geomorphologists of the past were often experts at field sketches and adept at turning them into explanatory cross-sectional diagrams. Looking at coral reefs, river valleys, and Alpine landscapes provided early landscape detectives with a key question – what could have formed these features?

Visual observation and recording from a viewpoint, however, do not permit detailed insights into the sediments and rocks which make up the landscape, and it is here that the great landscape detectives of the past made use of developing industrial technology. During the 18th and 19th centuries, in Britain and other countries, vast alterations and extensions were made to the transport routes, as canal and rail networks were opened up. Digging canals necessitated cutting into the landscape – dissecting it, in a sense. This allowed the landscape detectives greater access to its internal workings, providing them with the beginnings of a three-dimensional description of the landscape. Quarrying and other forms of mineral extraction gave similar glimpses under the Earth's surface, allowing new questions to be asked about familiar landscapes. In many cases, the view obtained by cutting into a landscape illustrated a complex palimpsest, cutting through many key episodes in the history of that landscape. For example, cuttings into valley side slopes may reveal ancient river sediments and illustrate how the fluvial system must have eroded into the landscape over time. As well as using these developments to advance their knowledge of landscape, people with geomorphological skills were often at the forefront of prospecting and evaluating new terrain. In the USA, during the late 19th-century expansion westwards, for example, there was a huge need for an understanding of landscape in order to assess what resources might be present, as well as what obstacles there might be to river navigation and the spread of railroads. Some of the earliest professional geomorphologists, including John Wesley Powell and W. J. McGee, were employed by the United States Geological Survey (USGS) during this period. There is at least some evidence to suggest they were the first to coin the term 'geomorphology'.

Geomorphologists also used, and continue to use, another method of visualizing how landscapes have evolved, that is, trying to observe a developmental sequence in time portrayed as an array of landforms in space. Imagine, for example, standing near the snout of a glacier that has been observed to be retreating over the past few decades. In front of most glaciers, you see deposits of sediment which have eroded and pushed down the valley as the glacier advances. Once a glacier starts to retreat, it loses the energy to carry this debris and deposits it as what geomorphologists call a 'terminal moraine'. In front of many large glaciers suffering from net retreat, we can see arrays of such terminal moraines reflecting different positions of the snout of the glacier over time. The great genius of some early landscape detectives was to realize that these spatial arrays could reflect a temporal sequence. In this case, the terminal moraine furthest from the current snout position would be the oldest, and so on until the newest, which would be nearest the snout.

The usefulness of substitution of space for time relies on two key facts which the early geomorphologists had to establish. First, whilst the net state of the glacier in this case must be retreating, it must also have experienced short periods of advance in order to create new moraines. Second, the sequence may only be a partial record of the recent history of the glacier. If advance dominates over a long enough period then older moraines will become bulldozed by the glacier, thus erasing part of its history. Observing moraines of different ages in some detail allows geomorphologists to build up a picture of how individual moraines evolve and develop over time. What we are noticing here is that however we look at the history of a landscape, directly or using space for time substitution, it is almost impossible for geomorphologists to separate out pure description from explanation – as soon as we start to describe, we are forced to interpret, and vice versa. This tells us something about the scientific method in geomorphology, and the links between theory and observations, to which we will return later.

A lovely example of substituting space for time in describing and interpreting an unfamiliar landscape is the work of Charles Darwin (1809–82) on coral reefs (Figure 4). During his time on the *Beagle*, he became the world's most well-travelled naturalist, and probably saw a wider array of landscapes than any other person. In the Pacific Ocean, which Darwin visited on the voyage of the *Beagle* in the 1830s, there are a number of different types of coral reef. Darwin identified three main types. A fringing reef is one that lies close to the shore of some continent or island. It consists of an uneven, generally flat, narrow platform about the level of low water. Between it and the mainland there may be a narrow channel or lagoon. A barrier reef occurs where there is a wider, deeper lagoon, and the reef lies at some distance from the shore, rising from deep water. An atoll is a reef in the form of a ring or horseshoe with a lagoon in the centre. Darwin's theory to explain these three types was based on the idea that subsidence had occurred. He argued that a succession from one coral type to another could be achieved by the upward growth of coral from a sinking platform, and that there would be a progression through time from a fringing reef, through the barrier reef stage, until, with the disappearance through subsidence of the central island, only a reef-enclosed lagoon – the atoll – would survive. Many years after Darwin put forward this simple but ingenious hypothesis, deep boreholes were drilled through some of the Pacific atolls, passing through more than 1,000 metres of coral sediment before reaching the basalt substratum of the ocean floor. This indicated that the coral had been growing upward for tens of

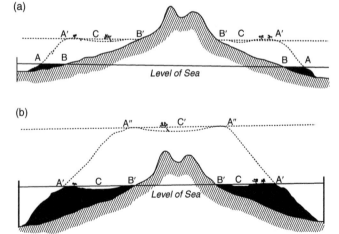

4. Charles Darwin's diagram of the evolution of coral reefs

millions of years as the crust subsided, thereby proving that Darwin's idea was basically correct.

Interestingly, Darwin developed his theory of coral reefs before he had seen one in the field, but he was able to apply his own observations from other settings and the descriptions of reefs made by others to try and solve the problem of their origin. As he puts it:

> No other work of mine was begun in so deductive a spirit as this, for the whole theory was thought out on the west coast of South America, before I had seen a true coral-reef. I had therefore only to verify and extend my views by a careful examination of living reefs. But it should be observed that I had during the two previous years been incessantly attending to the effects on the shores of South America of the intermittent elevation of the land, together with the denudation and deposition of sediment. This necessarily led me to reflect much on the effects of subsidence, and it was easy to replace in imagination the continued deposition of sediment by the upward growth of corals. To do this was to form my theory of the formation of barrier-reefs and atolls.

Making such observations of the surface and subsurface characteristics of a landscape permitted the early landscape detectives to start explaining how it had developed or, as William Morris Davis put it in 1904, how 'the geography of today is nothing more nor less than a thin section at the top of geology, cut across the grain of time'. The most obvious starting point to explain the past is to look around us at the present and see if we can find evidence of processes and changes occurring that we could then extrapolate back into the past. For example, if we want to understand how river valleys developed, we could look at the present processes operating within a range of river valleys and think about how they could produce the landforms that we also observe. Immediately, we can see the difficulties that the early landscape detectives faced. Looking at most landscapes, especially before the advent of specialized techniques which can now monitor the rates of processes, nothing much appears to be happening! Earth surface processes, as we have seen in Chapter 1, often occur very slowly. Many landscapes appear unchanging, or at least to be changing extremely slowly. The landscape detective was faced with two choices. Either they could assume that the processes they could see happening today had always been in place and thus the landscape had evolved incredibly slowly, or they could invoke something more dramatic. That 'something more dramatic' could be fast-operating processes known to occur in other places in the world today, but for some reason not happening in the landscape under study, or they could be something without an analogue in today's world – like a giant flood or a meteorite impact.

Up until the latter part of the 19th century, in Britain and other countries dominated by Christian religious beliefs, this led to a series of problems. According to contemporary Christian doctrine, the Earth had been formed a few thousand years ago by God and shaped by Noah's flood and other major events. Many landscape detectives at the time realized that this narrow interpretation of the biblical record was incompatible with the landscapes that they were trying to understand.

One of the key doctrines to emerge in the 19th century as applied to explanations of the natural environment is that of 'uniformitarianism'. Simply put, this states that the 'present is the key to the past' or, to put it another way, is based on the assumption that the processes that we can see around us operating on the Earth today are those that were responsible for past changes. We don't need to invoke any exceptional processes that we cannot see somewhere in the world today. During the 19th century, this doctrine was used in opposition to that of 'catastrophism', which asserted that several key parts of the global landscape had been created by supernatural forces or God-driven processes (such as Noah's flood). Uniformitarianism does allow rates of operation of processes to have varied in the past, but not their essential nature.

James Hutton (1726–97) is regarded by many as the founder of modern geology as a result of his ideas on the first published in 1788 as *Theory of the Earth: or, An Investigation of the Laws Observable in the Composition, Dissolution and Restoration of Land Upon the Globe*. He is also often regarded as a father of geomorphology because his theory of the Earth illustrated the importance of denudation (alongside uplift and oceanic sedimentation) in the development of the Earth's surface. Applying his knowledge of the slow present-day workings of fluvial systems to interpreting the development of large valley systems, he was forced to conclude that they were of immense antiquity, far older than permitted in the strict interpretation of biblical chronology accepted at the time. There was also, in his mind, no evidence for Noah's flood or 'diluvial' events having shaped the fluvial landscape, just these slow fluvial processes. In his words of 1788, 'there was no vestige of a beginning, – no prospect of an end', just slow, inexorable fluvial processes carving their way imperceptibly through the landscape. However ground-breaking Hutton's ideas were, he faced much uncertainty, as he had no real way of knowing the processes operating today had indeed operated at much the same rate in the past. Exogenic processes, as we explained in Chapter 1, are largely controlled in nature and rate by climate, and thus much explanation of landscape history requires an understanding of past climates.

Hutton's observations of slow but important denudation by rivers provided a key element of developing geomorphological theory. Towards the end of the 19th century, the role of rivers as agents of landscape development was more firmly established, thanks to the impressive (but now largely abandoned) ideas of William Morris Davis (see the box). Davis used many of the classical methods of landscape detection, focusing especially on visual observations and comparisons.

William Morris Davis

William Morris Davis (1850–1934) has been described as an Everest among geomorphologists. He was the leading American geomorphologist of the late 19th and early 20th centuries. He spent most of his career at Harvard University, where he was an exacting but skilful teacher. Above all, he was a very prolific author, writing more than 500 articles and books.

His great contribution was to produce a deductive model of landscape evolution, called the 'cycle of erosion', or the 'geographical cycle'. This was developed during the 1880s and 1890s. Cross-sectional diagrams and field sketches were important methods that Davis used to help visualize his cycle. Davis believed that landscapes were the products of three factors: structure (geological setting, rock character, and so on); process (weathering, erosion, and so on); and stage in an evolutionary sequence. Stage was what most interested him. He suggested that the starting point of the cycle was the uplift of a broadly flat, low-lying surface. This was followed by a phase he termed 'youth', when streams became established and started to cut down and to develop networks. Much of the original flat surface remained. In the phase he termed 'maturity', the stream valleys had widened so that the original flat surface was largely eroded away and streams drained the entire landscape. The streams began to meander across wide floodplains and the hillslopes gradually became less steep. In 'old age', the landscape became so denuded that a low relief surface close to sea level developed, with only low hills rising above it. This surface was called a 'peneplain'. Rejuvenation of a landscape and continuation of the cycle could occur as a result of tectonic uplift or other processes that lower the base level.

One of the major advances of 19th-century geomorphology and Earth history involved a very different hypothesis about the landscape-forming importance of fluvial erosion in Alpine and other mountainous environments. Here descriptions of landforms and sediments had revealed many enigmatic features – strange fossils which appeared to come from the sea but were found within high mountain environments, weirdly shaped valleys which looked as though they must have been carved by unfeasibly large rivers, boulders which had clearly been moved long distances by huge forces, and so on (Figure 5).

Invoking the slow, long-term operation of present processes clearly could not explain the landscape. Two major theories were developed. First, Noah's flood was proposed as the agent that had carved the valleys and deposited the marine fossils (which became known as the 'diluvial theory'). Second, it was proposed that in the past, glaciers had expanded hugely and that it was the power of ice, not water, which had carved the dramatic valleys and other unusual topographies. Louis Agassiz (see the box) has

5. Erratic boulder forming part of a lateral moraine overlooking the Mer de Glace, near Chamonix, France

become identified as the major proponent of this theory, and ultimately of the discovery that past ice ages were major transformers of the landscape. The glacial theory could not explain the marine fossils, and here again geologists were forced to accept that the biblical dating of the Earth could not be right, and that these fossils dated back to much earlier in the landscape's history when rocks were laid down before they were contorted into mountains. Thus, during the 19th century, we start to see a clear separation of geology and geomorphology as questions about the landscape become detached from questions about the underlying building-blocks.

So how did these early landscape detectives start to identify which processes could have operated in the past in order to test some of these ideas about landscape histories? One of the most obvious methods, still utilized greatly today, is analogy with other contemporary environments. For example, when trying to determine whether ice really could have carved the great glaciated valleys, arêtes, and pyramidal peaks observed in the Alps and the English Lake District, Agassiz and others looked at the recently exposed glacial erosion features surrounding valley glaciers within the Alps. Another method, also used commonly today, is that of experimentation. Some of the greatest geomorphological problems have been solved by carrying out experiments – such as Grove Karl Gilbert's work to simulate the production of meteorite craters, and Charles Darwin's experimentation with earthworms to calculate rates of soil turnover and erosion carried out by these humble creatures. G. K. Gilbert, a key figure in the history of geomorphology (as discussed in the box), produced craters in clay and sand targets by dropping clay bullets and shooting bullets into them and used the results to infer how craters on the surface of Earth and other planets developed. Ultimately, both approaches assisted the landscape detectives in building up plausible links between processes and resultant landforms. If these processes could then be related to specific climatic conditions, and if their rates of operation could be roughly quantified, then it became possible

Louis Agassiz

Louis Agassiz (1807–73) was born in Switzerland and died in America. He trained initially as a medic in Switzerland and Germany, but then went to Paris, where he fell under the tutelage of Alexander von Humboldt and Georges Cuvier who converted him to a twin career in zoology and geology. In the late 1830s, building upon the ideas of such figures as Ignatz Venetz and Jean de Charpentier, he developed the idea that the Earth had undergone an 'ice age' and that a huge ice sheet had developed over the Alps. He recognized the criteria for the glacial modification of landscapes, including moraines, erratic blocks, and the grooving and striation of rock. In 1840, he visited the British Isles and found that here too there was abundant evidence of former glacial activity. In 1846, he moved to America, and as soon as he landed in Halifax, Nova Scotia, he sprang onshore and was immediately met by 'the familiar signs, the polished surfaces, the furrows and scratches, the line-engraving of the glacier so well known in the Old World'. Such was his enthusiasm for the ice age that when he visited Brazil in the 1860s, he claimed to find glacial drift there as well! Not only did Agassiz identify many of the criteria for glaciation, he also explained many phenomena that had previously been attributed to Noah's flood, and made a fundamental contribution by recognizing that severe climate change had a major role to play in creating the landscapes we see today.

to develop some ideas of how, and how quickly, a landscape had developed and over what timespan.

In many cases, explanations of how landscapes developed were based on an intimate and quite complex relationship between observations and the development of a theory. There are three main things that help us explain landscapes: knowledge of the laws of nature, the potential causes of change, and the effects of those

Grove Karl Gilbert

Grove Karl Gilbert (1843–1918) was a remarkable American geomorphologist who, in many respects, was ahead of his time. Although he died over 90 years ago, his career exemplifies *par excellence* many of the concerns of modern geomorphology. Spending much of his career in the American West, he made diverse and impressive contributions to the discipline. He helped to explain and name the structure and topography of the Basin and Range province, with its many alternations of mountains and playas; he explained and classified the igneous intrusions that had created the Henry Mountains of the Colorado Plateau; he studied the greatest pluvial lake of the American West – Lake Bonneville – and recorded the evidence of its fluctuating levels; he established that large lakes could depress the Earth's crust and so contributed to the growth of ideas about crustal mobility; he helped to demonstrate that the craters on the Moon were the result of meteorite impact; he used laboratory flumes to carry out experiments on fluvial processes, and studied the environmental effects (siltation and so on) of hydraulic mining for gold. A great exponent of the use of hypothesis testing as scientific method, he was, as his biographer Stephen J. Pyne remarked, 'A great engine of research'.

changes as seen in the landscape. Different approaches in geomorphology use knowledge about two of those to derive information about the third (as shown in Figure 6). For example, the inductive approach uses observations of both causes and effects to try and develop theories and laws. Deduction, in comparison, involves knowledge of causes and the laws of nature to try and work out what the effects or outcomes are. Charles Darwin's work on coral reefs can be seen to come under this deductive category, as does W. M. Davis's cycle of erosion. Scientists are often taught that deductive and inductive approaches are the only two ways to approach science. However,

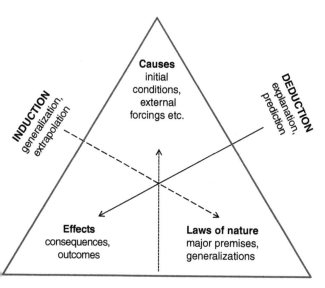

ABDUCTION
reconstruction, inference

6. Modes of explanation in geomorphology

many geomorphologists (both the early landscape detectives and geomorphologists today) use a third approach to scientific explanation, called 'abduction'. The basic approach here is to look at the end results, or 'effects' (a landform or landscape), and to use one's knowledge of how the world works (laws of nature) to work out what the causes have been. Whilst it is impossible to know how the great early landscape detectives actually did their science, the abductive method certainly provides a good basic explanation of how they interweaved observations and ideas to produce important insights into landscape histories.

A final part of the landscape detectives' toolkit in the 18th and 19th centuries was the ability to date events in the past. This dating has two important aspects: first, to say when a particular event

27

occurred at one location; and second, to correlate this with other events in different places. A major part of attempts to understand the ice ages was to date when ice advances occurred within the Alps and whether similar changes occurred in other places at around the same time. If large swathes of upland terrain within the northern hemisphere experienced glaciations at the same time, then one might search for very different causes of that glacial expansion than if only small areas were affected at any one time. Dating of landscape features was extremely difficult until the advent of radiometric techniques, which we will discuss in more detail in Chapter 3. However, stratigraphic position (that is, where a deposit occurs within a vertical sequence) was used as a relative dating technique from quite early on in the history of the geosciences, based on the Law of Superposition (which states that newer deposits will be laid down on older ones). Similarly, identification of fossils within certain strata also helped to identify individual deposits and correlate them across large areas. For example, understanding of the events of the Jurassic period was enhanced by identification of fossils such as ammonites in particular strata at locations known to date from the Jurassic. Finding ammonites in other locations helped to discern that these rocks too dated from the Jurassic period. Over more geomorphologically relevant timescales, identification of fossils and deposits thought to be diagnostic of particular glacial or interglacial periods helped to establish regional chronologies of glaciations during the ice ages.

How did these early landscape detectives know that they were right? Why do we value and celebrate some successful theories such as those of James Hutton, Louis Agassiz, and Charles Darwin, whilst many others have become forgotten or laughed at? Why has the work of a 'geomorphological giant' such as W. M. Davis fallen from favour today? Publishing is a highly important component of any science, and work on understanding landscapes is no exception. All of these scholars wrote copious amounts. However, scientific ideas need to be successful over the long term, as well as published at the time. A successful theory provides a good explanation of the particular phenomenon under examination, but it usually goes much further than that. It usually also provides a better explanation than other theories, or fits in with other theories and ideas, and also often gives insights into other phenomena. Darwin's theory of coral reef development, for example, fitted the observed facts better than any others. It has also stood the test of time, even when new ideas (such as the great idea of plate tectonics which came about in the 20th century) arose which felled a good many other theories. Similarly, Louis Agassiz's ideas on the ice age, whilst having been refined and complicated by observations from other places and more detailed chronologies, proved capable of explaining landscapes in many different places, and in turn were validated by Milutin Milankovitch's identification of the major motivating force for large-scale glaciations, that is, the changing orbit of the Earth around the Sun. James Hutton's ideas on the slow geological cycle in which land was created and denuded over vast timescales as a result of natural forces, not supernatural ones, paved the way for the evolution of modern geomorphological thinking, even though many of his ideas have proved in detail to be wrong or inadequate. W. M. Davis's theory was extremely powerful for a long time in geomorphology, but was ultimately superseded by new datasets and the advent of plate tectonics theory.

We leave this discussion of the early 18th- and 19th-century landscape detectives by thinking about how they link to geomorphology today in the early part of the 21st century. The intellectual bravery of a relatively small group of men (afforced by an even smaller group of women) ensured that some really big questions about how landscapes have developed and over what sort of timespans had been at least initially tackled by the dawn of the 20th century. In a very real sense, they set the agenda for our current geomorphology. But there were several key ideas that underpin our modern conception of landscapes which were not yet available, most notably the theory of plate tectonics, which now provides an extra dimension to our explanation of the long-term development of landscapes and the whole global system. The growth of geomorphological science went hand in hand with the expansion of capitalism within northern Europe and as imperialism spread more widely. The timing is not accidental, and we can see similar correlations between economic development and advances in landscape science during the 20th century, as industrial and mercantile expansion led to demands for increased scientific knowledge about the natural world. Finally, the fact that individual landscape detectives ranged widely between geological, biological, and geomorphological pursuits illustrates the many linkages between these different fields within the 18th and 19th centuries. Such linkages are paralleled today, but in a rather different way, by the advent of 'Earth system science'.

Chapter 3
Studying landscapes today

Geomorphologists now have at their disposal a vast range of techniques with which to probe past and present landscapes in order to answer key questions about how the world works. Whilst the last chapter focused on how the early landscape detectives used their knowledge of the present to understand the past, we now look at how our new techniques and ideas have revolutionized understanding of how landscapes function and contribute to the global system. One key theoretical advance which came about during the 20th century was the development of plate tectonics as a theory which underpins all Earth and environmental science, including geomorphology. Understanding how Earth surface processes and landforms interact helps in the quest to provide more complete explanations of how land, ocean, and atmospheric systems function and change over time. Thus, geomorphology is (or should be) central to Earth system science. In this chapter, we firstly introduce plate tectonics and then explain in more detail what we mean by Earth system science. We then outline the major types of techniques available to geomorphologists and how they are used to provide more detail on landscapes past and present. We conclude by illustrating with the example of the Tibetan Plateau how geomorphological techniques contribute to answering big questions about past and present global environmental change.

The work of James Hutton, William Morris Davis, and other early landscape detectives took place before the advent of the plate tectonics 'revolution'. By the mid-1960s, Earth science had indeed been revolutionized, but the evolution of ideas goes back a long way. One of the first formal statements that the Earth's crust was mobile was put forward in the book *The Origin of Continents and Oceans* by Alfred Wegener, published in German in 1915 and in English in 1924. Wegener was in many ways an outsider to Earth science, having a PhD in astronomy and becoming a professor of meteorology and geophysics. He formulated a theory to explain how the position of the continents relative to the poles and to one another had changed dramatically over time, but lacked evidence of what was the motive force for these movements. His theory was highly controversial and generated much discussion amongst geologists, who saw that, as R. T. Chamberlin put it in 1928, 'if we are to believe Wegener's hypothesis we must forget everything which has been learned in the last 70 years and start all over again'. Over the course of the 20th century, advances in understanding of how the Earth's system works (especially knowledge of the Earth's magnetism), coupled with various observations (especially of movement of the sea floor), proved that the Earth's surface was, indeed, divided up into a series of plates which had moved, and continue to move. Acceptance of the theory of plate tectonics encouraged the development of Earth sciences as an integrated discipline, as it helped explain the long-term development of climate, ocean currents, vegetation, and landscapes across the world.

The term 'Earth system science' originally came from NASA and is often abbreviated as ESS. One definition, adopted by the Earth System Science Partnership, is that:

Earth system science is the study of the Earth System, with an emphasis on observing, understanding and predicting global environmental changes involving interactions between land, atmosphere, water, ice, biosphere, societies, technologies and economies.

They define the Earth system as:

> ...the unified set of physical, chemical, biological and social components and interactions that together determine the state and dynamics of Planet Earth, including its biota and its human occupants.

Earth system science has, on the one hand, been celebrated as bringing a new holism to the Earth and environmental sciences, and as a major step forward; whilst on the other hand, it has been criticized as being a politically motivated attempt to show the importance of Earth observation which offers nothing new. So where does landscape fit into Earth system science? Taking the view of landscape as a palimpsest of rocks, soils, vegetation, animals, and human constructions, as we proposed in Chapter 1, we can see that landscapes (at all scales from the local to the global) are basic to ESS. What is less clear is how the local contributes to the global, or how individual landscapes are linked into global environmental change. Making these links is an important challenge to geomorphologists today, and to achieve this they need to utilize a battery of techniques.

The techniques available to landscape detectives during the 18th, 19th, and early 20th centuries were fairly basic. Whilst a whole array of field portable techniques permitted accurate survey of the shape of the land, it was much more difficult to obtain information about the constituents, ages, and wider context of the landscape. Starting with the advent of aerial photography, however, an explosion of techniques occurred, allowing better field measurements, dating, laboratory experimentation, remote sensing, and modelling of geomorphological data. What makes today's geomorphology exciting is that now these techniques are being used in an increasingly integrated fashion, alongside those of ecologists, climatologists, hydrologists, and geologists, to provide a comprehensive assessment of landscapes past and present.

Whilst field-survey techniques have been used routinely in geomorphological investigations since the days of Alexander von Humboldt, the advent of GPS-based systems has allowed much more rapid and precise data collection on topography and its change over time. Differential GPS (or DGPS, whereby information is collected about the location of a small roving GPS receiver, often sticking from a surveyor's backpack, relative to that of a base station) is now commonly used by geomorphologists to provide accurate topographical surveys of features such as sand dunes, beaches, and failure-prone valley side slopes. Resurveys can be carried out after some time and are used to evaluate rates and patterns of landform change. Such methods greatly increase our level of understanding of landform change, and bring much more sophistication and accuracy than previously used methods such as resurveying the position of one or two stakes or erosion pins relative to a mobile landform. Ground-based LIDAR (light detection and ranging) techniques provide an even quicker and highly accurate method of surveying detailed, and particularly inaccessible, topography, such as rocky cliffs. LIDAR equipment sends out a laser beam which scans across a surface, and then measures the time it takes for the laser beam to bounce back and the way it gets deflected by the surface under survey. This effectively and precisely maps the surface topography in a non-contact and quick way. Photogrammetry can be used in a similar way to map a surface in detail and, with the use of fixed reference points, to document change over time. Such techniques can now provide geomorphologists with extremely detailed portrayals of how the landscape surface changes as a result of erosion, deposition, and other Earth surface processes. Whilst many of these techniques remain highly expensive, bulky, and reliant on a power source such as a generator, the technology is improving rapidly, and more and more geomorphologists are taking the opportunity of using it.

Through precise and accurate survey and resurvey, we can now see in much more detail how landforms change. The 'promiscuous barchans' (more prosaically described as shifting crescentic sand dunes) which so seduced the British geomorphologist Ralph Bagnold in the Western Desert of Egypt, for example, have now been surveyed and resurveyed in great detail using DGPS in comparison with earlier air photograph surveys. We now have a much greater idea of how they move and change over short and longer timescales. Similarly, combinations of aerial LIDAR and air photography have revealed how coastal barrier islands change in form and location as a result of both short-term forces (such as storms and hurricanes) and longer-term change (such as sea level rise over the past few decades).

A second set of field techniques is now being used by geomorphologists to probe more deeply under the surface of the landscape. Such geophysical techniques allow a two- or even three-dimensional view of the materials and structures which make up the landscape. Two techniques are particularly well utilized by geomorphologists: ground-penetrating radar (or GPR), and a suite of geoelectrics techniques (particularly resistivity surveys). GPR has proved to be an exceptionally useful tool for investigating the layered structures of large sand dunes. Electromagnetic pulses are sent from a transmitter antenna into the ground and the time taken for them to reach a receptor antenna recorded. The time taken for the signal to return depends on the ground conditions. Moving the antennae in a transect or grid across a landform, for example, allows the operators to build up a two-dimensional subsurface picture. Some sediment layers reflect the radar signal more clearly than others. Resistivity surveys work on a similar principle but utilize the response of the materials to a current applied between two electrodes on the surface. Some materials conduct electricity better than others, and thus the resistance recorded at the receptor electrode varies depending on the material's characteristics. Resistivity methods are particularly helpful in searching for voids within a landscape (for example, looking for caves within a karst landscape) and in looking at variations in moisture within a deposit (as water content is a key control over the resistance of a porous material). Increasingly, geomorphologists use these two techniques in an integrated way, as the information provided by each of them is complementary and adds to the explanation.

GPR and resistivity surveys have been used together, for example, to investigate scree slopes within mountainous terrain. Between the two, they reveal much detail of the internal structure of scree deposits, which can be used to infer how they have developed – for example, whether they have formed in one or two large pulses (perhaps as a result of specific climatic fluctuations) or in a more diffuse way as a response to slow, nearly constant weathering of the upper rock faces. GPR has also been used to probe deeply into large linear dune features within the Namib Desert, Namibia. Changing patterns of sand layering have been revealed, illustrating a complex history of deposition and erosion.

Geomorphologists today make much more use of the laboratory than did the early landscape detectives. Two main types of lab investigation are carried out: diagnostic analyses (of types of material, their ages, and any palaeoenvironmental signals they might contain) and experiments. Dating techniques are by far the most important diagnostic analysis used by geomorphologists. The key attribute of most modern-day dating techniques is that they are absolute (not relative) and radiometric – that is, based on the differential decay of radioactive isotopes over time. The best-known radiometric technique, used widely in the Earth and environmental sciences, is ^{14}C (also known as carbon 14, radiocarbon, or carbon dating). Carbon dating can be used to date any organic carbon, such as skeletal material, charcoal, or plant remains. It has become a highly refined and accurate technique over recent years, but is not necessarily the technique of choice for many geomorphological studies, for two main reasons – the timescale over which it can be applied (up to around 70,000 years at most), and the fact that it cannot be used on most geomorphological sediments (but only on carbon-rich organic materials).

Two other radiometric dating techniques have become particularly important for today's geomorphology as they can be used on inorganic sediments and cover longer timescales. The first of these is optically stimulated luminescence (or OSL), which works particularly well on quartz (which is a highly resilient mineral and thus makes up a large fraction of sediments around the world). Whilst buried in a deposit, quartz and other minerals receive radiation from their surroundings. This radiation gradually causes changes to quartz atoms, as electrons are displaced. The 'signal' builds up in a predictable way over time (depending on the radiation 'dose rate' received in that area) until the quartz grain becomes exposed at the surface. When, as a result of natural processes, the sediment grains become exposed to sunlight (called bleaching) the 'signal' is re-set. These phenomena provide us with an opportunity to date the exposure or burial of sediments. In order to do this, a sample is taken under darkness, and in the lab bleached artificially and the amount of stored signal released during this bleaching is measured. This allows the length of time that the sediment has been buried to be calculated (given a number of other bits of information about the landscape, such as the local dose rate, often measured in the field with a gamma spectrometer, as shown in Figure 7). Dating over several hundreds of thousands of years can be obtained using this method where the environmental dose rate is slow. Examples of the geomorphological use of OSL dating include investigating vertical cores taken through desert sand dunes, loess deposits (wind-blown dust), or fluvial

7. **Using a gamma spectrometer to measure environmental dose rate as part of OSL dating**

terraces to look at the ages of different layers, and using this information to build up a chronology of deposition over large areas.

Cosmogenic dating works in a similar way (see the box). It is often used to date surface exposure and rates of denudation, but can also be used to look at burial histories. One great advantage of cosmogenic dating is the often very long timespans that it can cover. For example, dating of rocky surfaces on inselbergs (isolated rocky hills) within the Namib Desert in southern Africa have shown that denudation in this hyper-arid environment is extremely slow, amounting to something like 10 metres over 1 million years. Cosmogenic, OSL, and a suite of other absolute dating techniques are starting to provide geomorphologists with a really good chronology of long- and shorter-term landscape change.

Cosmogenic dating

The Earth is constantly being bombarded by cosmic rays. These induce nuclear reactions within the upper few metres of the Earth's surface, producing cosmogenic nuclides. With the development of advanced spectrometric techniques, it became possible from the 1980s to measure the concentrations of these nuclides. This enables dating of surfaces to be undertaken, as their nuclide concentrations are interpreted as reflecting the time elapsed since the surface was first exposed. The range of dates that can be obtained is impressive, as they can extend over several thousands to several millions of years. However, the method has its limitations. Exposure dating requires that the surface formed over a short time period. Examples of surfaces suitable for cosmogenic dating include fault scarps, lava landforms, landslides, meteorite impacts, catastrophic floods, wave-cut platforms, glacially eroded bedrock, and erratic boulders. Another requirement is that the surface form must be preserved over the period of exposure, so one needs to date surfaces with very low rates of erosion.[10] Be and ^{26}Al are the two nuclides most often used.

No technique is without its problems, however, ranging from sheer expense to a whole host of factors which can generate erroneous dates. Furthermore, whilst we can be ever more precise and accurate about dating individual sediments, big questions still remain about what these dates mean in terms of how and why the landscape is changing. As an example, we can now create a vertical profile of dates every 50 centimetres or so down through a large linear desert dune. But what do these dates really tell us about how conditions of aridity, windiness, and sand supply have changed over the timescale measured? Does the lack of sand dating from a particular time period mean that there was no sand deposited then? And, if so, why not? Was it too wet, not windy enough, or was there not enough sand blowing around? Or does the lack of sand mean that it was subsequently eroded as a result of small-scale alterations in dune topography before the next layer was deposited? We now have extremely detailed records of the history of many sedimentary archives, but still much uncertainty remains over what they actually mean. This is a really good example of where geomorphologists working on current process-form questions within today's landscape can contribute much to questions of how past landscapes have evolved.

Experimentation, both under controlled conditions in the lab and under real field conditions, is another key technique utilized by geomorphologists today, especially in the quest to understand how landscape processes operate and produce change in landforms. In the lab, for example, it is possible to simulate diurnal and other cycles, and to subject rock and sediment samples to them in order to assess their susceptibility to a range of processes of weathering and erosion. Large blocks of chalk, for example, have been placed into freezing rooms in order to simulate the effects of permafrost on chalk landscapes, whilst other rocks have been placed in wind tunnels and blasted with sand to simulate aeolian (of the wind) abrasion. Such experiments have the great advantage that they both simplify and accelerate natural conditions, thus allowing relatively quick assessments of change. Their disadvantage, of course, is that they are not natural. Field experiments can also provide useful information on landscape dynamics, in more realistic settings. Portable rainfall simulators and wind tunnels, for example, have been taken into the field in order to assess the rate of soil erosion under differing conditions. In these cases, the materials under study are entirely natural and undisturbed, whilst the environmental conditions are simulated. Another approach to field experimentation involves exposing prepared materials (such as blocks of different rock types, sometimes pre-treated to simulate past erosion and weathering) to natural climatic conditions. These experiments usually take much longer to carry out, requiring years or decades to produce any noticeable change.

Starting with air photography, geomorphologists have made extensive use of the ever-growing number of remotely sensed data techniques available. We can now see large-scale lineaments and features within the landscape from Landsat and other satellite imagery, which were simply not possible to see by any other means. Freely available Google Earth imagery provides an unparalleled view of many parts of the land surface which were previously only available at large expense, if at all. The resolution of satellite imagery is becoming ever more impressive, with Quickbird and other satellites providing metre- or even centimetre-scale resolution, although this is still generally expensive to obtain. Satellite imagery is particularly good for showing the look of terrain in inhospitable places which could not easily be studied in the field, as well as for identifying features which cannot easily be seen on the ground. For example, within many parts of the UK, there are subtle topographic signatures of past processes such as periglaciation, whereby during glacial periods ice within soils in areas not directly subjected to glaciations caused deformation to the sediments. Such features are hard to spot, and have previously largely been identified in cross-section in quarries and cuttings, but show up well on many Google Earth images as ghostly patterns within crops.

Remote sensing is not just about satellites, however, as airborne LIDAR, photography, and videography also contribute to today's geomorphological toolbox, especially for studying change along sedimentary coasts. Remote sensing is also not just about images, as there is much useful data collected from radar and other wavelengths. Satellite Radar Interferometry (or SRI), for example, is a technique of growing importance for studying geomorphic change, especially in the form of vertical displacement as might be associated with mass movements or tectonic deformations. Radar (the term comes from the acronym for radio detection and ranging) commonly uses microwaves nowadays to investigate topography (and works in the same way as LIDAR, as discussed above). Interferometry involves the overlying of two radar datasets collected at different times to quantify and locate changes in surface height relative to the satellite. SRI has been used to visualize the bulging in the sides of a volcano prior to an eruption and to produce topographical datasets for large areas. In the Shuttle Topography Radar Mission, the space shuttle Endeavour was used to obtain elevation data on a near-global scale from a specially modified radar system during an 11-day mission in February 2000. The data are available for much of the world at 90-metre resolution (30-metre for the USA).

A final, and hugely important, technique is that of modelling. By 'modelling', we mean the simulation of landscape processes and change. Such modelling is routinely carried out on computer, but can also involve experimental simulations, or the development of conceptual models. Such conceptual models are nothing new, as many of the early landscape detectives used sequences of diagrams to build a conceptual model of change over time. What is most exciting today is that now such modelling can be quantitative rather than qualitative, and that it is increasingly linked to data collected using the array of field, lab, and remote-sensing techniques introduced above. We can now provide a highly integrated view of landforms and landscape change using these techniques. We can also now provide a view of the landscape which ranges widely across the scales – from the short-term and micro-scale assessment of sand grain impacts on rock surfaces, to the long-term and macro-scale identification of the evolution of whole landscapes.

The Tibetan Plateau

Let us now look at a very large and important element of the global landscape, the Tibetan Plateau, which exemplifies well how geomorphologists, along with climatologists, palaeo-ecologists, and geologists, contribute to understanding some really big questions within Earth science. The Tibetan Plateau covers an area of around 3 million square kilometres (half the size of the USA) and has an average altitude of over 4,500 metres. Together with the Himalayan-Karakoram mountains, which make up its southern rim, it contains almost all of the Earth's landscape over 4,000 metres in height, as well as including the majority of the world's peaks over 7,000 metres. The origin of the Himalayas and the Tibetan Plateau lies in the most dramatic tectonic event to affect the Earth within the past 50 million years. As India broke away from Gondwana as a result of plate tectonics, it drifted northwards from a position about 20–40°S of the Equator at 70 million years ago, to a position between 10°N and 10°S some 40 million years ago. As it did so, the Tethys Oceanic Plate, which lay between India and Eurasia, was subducted beneath the Eurasian Plate, a process which ceased around 40–50 million years ago. It was at this time that India collided with Asia for the first time. However, India kept on ploughing into Eurasia at a very fast rate, producing some extremely high rates of uplift over long periods of time. As we saw in Chapter 1, the altitude of land surfaces is determined largely by the balance between uplift and erosion, and these are largely controlled in turn by tectonics and climate. According to the latest scientific evidence, the Tibetan Plateau (or at least the southern portion of it) appears to have stabilized in altitude since about 15 million years ago (mid-Miocene), but it has been an important element of global topography for much longer.

What impacts did this tremendous plate tectonic collision have on regional and global landscapes? Looking first at the regional scale, uplift over such a large area caused aridification and cooling of central Asia as well as the onset of the Asian monsoon. Rapid and extensive uplift produces large mountains, which influence the climate by providing a barrier to large-scale, high-level air movements. We know that the Asian monsoon system developed as a direct consequence of the uplift of the Tibetan Plateau and Himalayas. This, in turn, produced accelerated erosion over many parts of the Himalayas (as, broadly speaking, more rain equals more fluvial erosion), which then affected rates and patterns of uplift. We can see here the essential interlinkages between the tectonic and climatic changes in the development of many huge river systems, including many spectacular gorges, within the Himalayas. Mass movements are also very active in areas with steep slopes, high rainfalls, and tectonic activity – and thus the Himalayan region has also been eroded by mass movements, whose debris ultimately becomes removed within the river systems. Furthermore, the rise of the Tibetan Plateau produced the largest area of glaciated terrain outside the polar areas (Figure 8), with knock-on impacts on the world's hydrological cycle.

Events within the Tibetan Plateau area also had, and continue to have, much wider ramifications for the global climate. Uplift over large areas causes an increase in rock weathering, and this in turn produces a reduction in global carbon dioxide in the atmosphere and global cooling. The rise of the Tibetan Plateau and the Himalayas, and the high erosion rates by glaciers and

8. Part of the Tibetan Plateau, as seen from space

rivers, exposed a vast amount of highly weatherable material. The rocks involved in the collision zone are dominantly silicate in composition, and the chemical weathering of silicate minerals when considered over long (tens of millions of years) timescales is known to contribute to removal of carbon dioxide from the atmosphere (known as 'drawdown'). This drawdown of carbon dioxide contributes to the global carbon cycle and, in turn, to global climate. Less carbon dioxide in the atmosphere leads to colder global conditions and vice versa. However, global climatic changes have also had an impact in turn on the Tibetan Plateau landscape. At the Eocene-Oligocene transition around 34 million years ago, for example, evidence suggests that the onset of arid conditions on the Tibetan Plateau was caused by global cooling. These arid conditions will have impacted upon erosion rates and affected the balance between erosion and uplift. Increasing aridity can produce both decreased erosion (less water to produce erosion via runoff) and increased erosion (less water leads to reduced vegetation cover and therefore surfaces become more prone to erosion). Arguments continue amongst scientists over which effect may dominate. Thus, there are complex two-way feedbacks between tectonic uplift and climate.

One of the major challenges for Earth scientists over the past couple of decades has been to try to decipher evidence in the Tibetan and Himalayan landscapes for these various processes and feedbacks. In this regard, geomorphological evidence has often been crucial, alongside techniques such as thermochronological dating techniques, pollen analysis, and high-resolution GPS survey. Many geomorphologists have been involved in attempts to quantify the operation of denudational processes to help test some of the theories about the interrelationships between the rise of the Tibetan Plateau and climate change. For example, evidence of the intensity of weathering at different time periods has been inferred from detailed analysis of the geochemistry of ocean floor sediments which have been brought down in rivers draining the Himalayas. Geomorphologists have also played key roles in reconstructing glacial and fluvial erosion rates and patterns from remotely sensed digital elevation model data and large-scale geomorphological mapping and modelling. Such work utilizes new techniques to collect data, but is based on established understanding of how geomorphic systems operate and respond to climatic and tectonic forcing.

Whilst the early landscape detectives of the 18th and 19th centuries might find a lot of the scientific work being carried out on the Tibetan Plateau hard to understand because of the range and depth of new techniques used, they would certainly be pleased to see geomorphologists involved. The Tibetan Plateau illustrates how a range of landscape evidence (often collected remotely) can be used to try to tackle such large-scale and important questions as how major plate tectonic events influence global climate and vice versa. What is clear is that much is disputed today about such long-term and large-scale questions in Earth system science. Each discovery is accompanied by debate and discussion. The scale and complexity of the questions being asked are enormous, and despite the best available techniques being used, there is still a considerable lack of evidence over the long timespans involved and difficulty in interpreting the evidence that survives.

Chapter 4
Landscapes, tectonics, and climate

The landscapes of the Earth are hugely influenced by two major factors that have become the focus of much research in the Earth sciences over the last 50 years: plate tectonics and climate change. These factors, and their impacts on landscapes, are of such importance to geomorphology that we want to devote a whole chapter to them. Tectonic processes have both produced the major lineaments of the Earth's landscape today and also shaped the geology which underpins it. Climate changes, especially the broad sweeps of glacial and interglacial conditions during the Quaternary period, have similarly left a major imprint on the landscapes of today. As we have outlined earlier on, climate and tectonics are the two fundamental factors that condition landscapes. Growth in knowledge about tectonics and climate change has been extremely rapid since the middle of the 20th century, and we now have a good overall understanding of their impacts on Earth's landscapes over space and time. We review the state of the art of both plate tectonics and climate change and give some tangible examples of how they have shaped, and continue to shape, our landscapes.

In the 1960s, a new concept of the outer layers of the Earth developed that was to tie in with the venerable idea of continental drift. It was postulated that the Earth's crust and the upper part of the mantle (the rigid part called the lithosphere) consist of a set of rigid plates which rest on the weaker and deformable asthenosphere and underlie oceans, continents, or a combination of the two. Seven major plates and four or five minor plates are recognized, together with eight or nine 'platelets'. They appear to average about 100 kilometres in thickness, and the larger ones have areas of 65 million square kilometres. These plates move, and plate tectonics is the name given to the study of how they move and interact.

Many of the large-scale features of the Earth's surface, such as most volcanoes and mountain ranges, are associated with the boundaries between plates. Some plates spread apart along divergent junctions, typified by the large rift running down the crest of the Mid-Atlantic Ridge. This particular feature results from the contact between the American Plates, on the one hand, and the Eurasian and African Plates, on the other. Sea-floor spreading takes place along such a boundary, as the void between the receding plates is filled by molten, mobile material that rises from the interior of the Earth. The material solidifies in the crack, and the plates grow as they separate. The Atlantic Ocean did not exist 150 million years ago, and it opened out by this process, leading to the disruption of an ancient supercontinent – Pangaea.

The rate of divergence either side of the oceanic axial rift helps to explain the nature and distributions of islands and seamounts (submarine volcanoes). Along slowly spreading zones, such as those typical of the Atlantic, volcanic material tends to accumulate relatively close to the locus of spreading, so that there is a well-defined mid-oceanic ridge, with relatively massive mountainous islands like the Azores. However, where, as in the case of the Pacific, the rate of spreading is quite rapid, growing volcanic structures are moved large distances from the zone of origin, where they eventually sink and become dormant. The reason for such sinking is that oceanic crust contracts and becomes denser as it cools. Thus, on the floor of the Pacific, there are a very large number of seamounts, and some of them – called guyots – have flat tops because they were planed off by wave action before they sank. The process of divergence is most prevalent in the oceans, but there are two major zones of spreading among the continents, one in Africa and the other in Asia. The African zone (also known as the Great Rift Valley) extends from the Red Sea as a series of rifts through the East African highlands, and it is predicted that these rifts will eventually open further to form an ocean. These are the site of some major lake basins, including Lakes Malawi, Tanganyika, and Kivu. In the Asian continent, the site of spreading is in Siberia, where there is a rift system of which the deeper parts are occupied by Lake Baikal.

Plainly, however, if plates separate in one place, unless the Earth is expanding, they must come together somewhere else. This they do at convergent junctions, and the zone of contact is one that is dominated by crumpled mountain ranges, volcanoes, earthquakes, and ocean trenches. At the collision point, one plate normally plunges beneath the other, and the material of the submerging plate is gradually re-incorporated into the upper mantle and crust. The area where the material is lost is called the subduction zone. As the oceanic plate plunges beneath the continental plate with which it has collided, parts of it begin to melt and form magma, some of which reaches the surface as lava erupting from volcanic vents. Furthermore, as an oceanic plate is subducted, plunging down at an angle of between 30° and 60° into the Earth's interior, the surface of the crust is drawn down to produce one of the ocean trenches, and the movement is jerky, which leads to earthquakes.

The third type of boundary is called a transform or transcurrent boundary. This exists where two plates do not come directly into collision but slide past each other along a great fault. The most famous example of such a boundary is the San Andreas Fault in California, which separates the northward-moving Pacific Plate from the North American Plate. This too is a zone of intense earthquake activity. Earthquakes here play a major role in shaping the landscape, sometimes producing distinctive landforms such as fault scarps. Fault scarps are often visible, where tectonic movement along a fault has led to uplift of one side and/or downthrow of the other – producing a step in the hillside. In other circumstances, earthquakes also play a big role in Earth surface processes, for example acting to trigger off mass movements on many hillslopes.

Mountain-building starts with the formation of an ocean by sea-floor spreading. Thick muds and silts eroded from the land accumulate on the continental slopes alongside thinner limestones and sands on the continental shelves. After perhaps 100 or 200 million years, a trench forms on one side of the ocean and the ocean crust is subducted under the continent. This may be followed by the consumption of the sinking crust, leading to the formation of a chain of volcanoes called an island arc. After a further period, the island arc itself gets swept into the continents, and finally the ocean becomes closed and the continents collide. When this happens, the sediments of the island arc, together with those of the continental margin, are caught as if in a vice, and so folding and overthrusting takes place in a zone of crumpling, with slivers of oceanic crust caught up too. Furthermore, molten igneous rock, generated as the oceanic crust melts in the depths, rises into the crust and so promotes uplift. In this way, some of the great fold mountains are formed. The great Karakoram Mountains in Pakistan are composed of lava and the remains of an island arc that was squeezed into the Eurasian Plate as the Indian Plate moved northwards. This explains why the summit of Mount Rakaposhi (7,788 metres), one of the highest mountains in the world, is formed of oceanic crust material.

As well as the direct impacts outlined above, plate tectonics influences Earth's landscapes indirectly and at smaller scales through its control on the nature and arrangement of rock types in any one area. Rock type is a major factor in the nature of landscapes, and indeed some geomorphologists specialize in landforms and processes associated with particular rock types such as granites, sandstones, or limestones. Rock type is important as a control on geomorphology because it influences the rate of erosion. Some rocks are far more resistant to denudation than others. Over time, this produces differences in topography across large areas, as more susceptible rocks are eroded quicker than their surroundings. Rock type is also important because it determines the nature of weathering and erosion, and thus the types of relief produced. For example, some rocks with clear jointing patterns, or pronounced layering of hard and softer beds, tend to weather and erode into distinctive landforms, as the erosion becomes concentrated in particular zones.

Rocks and relief – the example of limestone landscapes

Limestones, composed primarily of calcium carbonate, are rocks in which the effects of chemical weathering, especially solutional attack, are particularly evident. Limestone dissolves in acidic waters, producing distinctive landscapes. Some of the world's most striking limestone landscapes are in southern China, where hot, wet climatic conditions have prevailed for a very long time, encouraging long-term dissolution. But limestones are extremely variable in their composition, hardness, jointing, and so on, so that their response to solutional modification is very varied. Thus, the soft Cretaceous chalk of southern England has very different scenery from some of the harder, denser, more crystalline, widely jointed Carboniferous limestones that form the classic limestone terrain of areas such as South Wales and north-west Yorkshire. Areas where the limestone is thick, massive, and extensive, and where the water table is at depth, may develop characteristic solutional features which in extreme cases produce a distinctive type of landscape called 'karst'. Karst areas often contain many surface solutional landforms such as limestone pavements with their clints (blocks) and enlarged joints (grikes). However, the typical landform of karst is the closed depression (dolines, poljes) through which surface water is carried underground. The evocative term 'swallow hole' is used in Yorkshire to describe such features. Closed depressions are often partly caused by collapse, but are also shaped by solution attacking joints. In many karst areas, such as the sinkhole plain, Kentucky, closed depressions form a network across the landscape. In wet tropical karst areas, they are often found in association with intervening residual hills, such as the dramatic cones and towers of southern China. Finally, karst landscapes are also characterized by extensive underground landforms, such as cave passages, associated with subterranean drainage systems.

It has been increasingly apparent that world environments have been subject to frequent and massive climatic changes during the course of the latest period of geological time – the Quaternary. Even in the last 20,000 years, the area of the Earth covered by glaciers has been reduced to one-third what it was at the glacial maximum; the waters thereby released have raised ocean levels by over a hundred metres; the land, unburdened from the weight of overlying ice, has locally risen by several hundred metres; permanently frozen ground and tundra conditions have retreated from extensive areas of Europe; desert sand fields have advanced and retreated; and inland lakes have flooded and shrunk.

The changes of climate have occurred over a wide range of timescales. The shorter-term changes include such events as the period of warming that took place in the first decades of the 20th century, and the years of low rainfall and high temperatures which contributed to the formation of the 'Dust Bowl' in the High Plains of the USA in the 1930s. But changes lasting hundreds of years were also characteristic of the last 10,000 years (a time often called the Holocene), including a period of glacial advance between about 1500 and 1850 called the 'Little Ice Age'. This cold spasm, besides causing glaciers to extend down their valleys, was also a period of severe avalanching, landsliding, rockfalls, and flooding in countries like Norway and Switzerland.

The fluctuations within the Pleistocene (the portion of the Quaternary that preceded the Holocene) consisted of major cold phases (called glacials) and warm phases (called interglacials) that lasted in total around 2 to 3 million years. The cooling that led up to the glacials of the Pleistocene is generally called the 'Cainozoic (or Cenozoic) climate decline'. During the Tertiary, which started at the end of the Cretaceous about 65 million years ago, temperatures showed a general tendency to fall in many parts of the world. Thus, in the North Atlantic region in the early Tertiary, conditions favoured a widespread, tropical, moist forest type of vegetation. Many rocks may have been deeply rotted by intense chemical weathering at that time. By Pliocene times, the degree of cooling was such that a more temperate flora was present in the North Atlantic region, and at just under 3 million years ago, glaciers started to develop in the mid-latitudes.

In the Quaternary, the gradual and uneven progression towards cooler conditions, which had characterized the Earth during the Tertiary, gave way to extraordinary climatic instability. Temperatures oscillated wildly from values similar to, or slightly higher than, today in interglacials to levels that were sufficiently cold to treble the volume of ice sheets on land during the glacials. Not only was the degree of change remarkable, but so also, according to evidence from the sedimentary record retrieved from deep-sea cores, was the frequency of change. In all, there have been about 17 glacial/interglacial cycles in the last 1.6 million years.

The last glacial cycle reached its peak about 20,000 years ago, with ice sheets extending over Scandinavia to the North German plain, over most of Britain (except the south), and over North America to 39°N. To the south of the Scandinavian ice sheet was a tundra steppe underlain by permafrost, and forest was relatively sparse to the north of the Mediterranean. In low latitudes, sand deserts were considerably expanded in comparison with today. Ice covered nearly one-third of the land area of the

Earth, but the additional ice-covered area in the last glacial was almost all in the northern hemisphere, with no more than about 3% in the southern. None the less, substantial ice developed over Patagonia and New Zealand. The thickness of the now-vanished ice sheets may have exceeded 4 kilometres, with typical depths of 2 to 3 kilometres. The total ice-covered area at a typical glacial maximum was 40 x 106 square kilometres, compared with the present 15 x 106 square kilometres. The temperature change that occurred over land was substantial. The presence of indicators of permafrost in southern Britain suggests a temperature depression of around 15°C during glacial phases. Mid-latitude areas probably witnessed a lesser decline – perhaps 4–8°C was the norm.

The glacials had a multitude of impacts on the landscape that are still visible today. Ice sheets caused considerable erosion and excavation, producing characteristic landform assemblages with cirques, arêtes, U-shaped valleys, roches moutonnées, and other forms. They also transformed drainage patterns, as the lake-studded landscapes of the Laurentian Shield in Canada and of Scandinavia testify. Elsewhere, they deposited boulder clay and outwash gravels. Beyond the glacial limit, fine particles blown from outwash plains settled to produce great belts of wind-blown silt – loess – in areas like Central Europe, China, New Zealand, and the Mississippi valley of the USA. Tundra conditions, with underlying permafrost, created great slope instability and the accelerated erosion of rivers, the evidence for which is still very apparent along the escarpments and valleys of southern Britain.

For many years, it was believed that the climatic changes associated with the glacials and interglacials also affected lower latitudes and that during glacials wetter conditions existed in the tropics and sub-tropics, causing lakes to reach high levels and rivers to flow in areas that are now arid. Such humid phases were called pluvials, and the warm, dry phases between them interpluvials. It was widely believed that the humid tropics were little affected by the changes of climate that transformed higher latitudes. There is indeed much evidence that at certain times in the past there was more water in desert areas; huge lakes, for example, filled the now largely dry basins of the south-west USA. However, there is also evidence that in other areas the periods of the glacials were times not of increased humidity, but of reduced precipitation. The most spectacular evidence for this is the great expansion that took place in the distribution of sand dunes in low latitudes. Dunes cannot develop to any great degree in continental interiors unless the vegetation cover is sparse enough to permit sand movement by the wind. If the rainfall is much over about 150 millimetres per annum, this is not possible. Studies of air photographs and satellite imagery indicate clearly that degraded ancient dunes, now covered in forest or savanna, are widespread in areas that are now quite moist (perhaps with annual rainfalls of the order of 750–1,500 millimetres).

The climato-vegetational changes of the Quaternary era were equalled in importance only by the worldwide changes in sea level that took place, though these themselves were caused partially by climatic factors. The sea level changes affected the configuration of coastlines, the size and existence of islands, the migration of plants, animals, and humans, and the degree of deposition and erosion carried out by rivers in response to a fluctuating base level. The effects of such changes can be seen along most shorelines. Where there are stranded beach deposits, coral reefs, shell beds, and platforms backed by steep cliff-like slopes, one has evidence of emerged shorelines. One also often has evidence of submerged coastal features such as the drowned mouths of river valleys (rias), submerged dune-chains, notches and benches in submarine topography, and remnants of forests or peat layers at or below present sea level. Many coasts show evidence of both emergent and submergent phases, such as parts of the southern coast of England where raised beaches and drowned estuaries are found in close proximity.

The most important cause of worldwide, or eustatic, sea level change in the Pleistocene was glacio-eustasy. When the ice sheets were three times as voluminous as today, a great quantity of water was stored up in them, and thus there was less water in the oceans. Sea levels may well have dropped between 100 and 170 metres, thereby exposing most of the world's continental shelves as dry land. The consequences of low glacial sea levels included the linking of Britain to the continent of Europe, of Ireland to Britain, of Australia to New Guinea, and of Japan to China. The floors of the Red Sea and the Persian Gulf were also dry land. In the Mediterranean, large plains existed off the coast of Tunisia and fringed most of Italy, southern France, eastern Spain, and much of Greece. The Aegean and Ionian islands were linked up to each other and to the mainland. Anatolian Turkey was connected to Europe by land-bridges across the Bosporus and the Dardanelles, while most of the Cyclades were merged into a single island.

The ice caps may have been slightly smaller during interglacials than they are today, and this may have caused sea levels to be a few metres higher than now, producing raised beaches in some coastal areas. In the Holocene, sea level rose very quickly, especially between 11,000 and 6,000 years ago. This transgression is often called the 'Flandrian Transgression'. It inundated the North Sea (flooding 'Doggerland' – an area that had been inhabited by Mesolithic peoples), broke the land link between Britain and Ireland, and drowned many river valleys to give features like the indented coastline of south-west England, with its winding rias. The same transgression cut off many islands that had previously been linked to the Mediterranean mainland and flooded the continental shelf between Australia and South-East Asia.

As well as glacio-eustatic changes in sea level, glacio-isostatic changes were also locally important during the Quaternary. During glacial phases, as the ice caps expanded, water loads were transferred from the oceanic 70% of the Earth's surface to the glaciated 5%. This led to local depression of the crust, producing isostatic downwarping of the continental margins, and thus leading to relative sea level rise. By contrast, the release of the weight of the ice resulting from melting in interglacials led to uplift of the land, and a relative drop in sea levels. In the areas that were covered by ice, the degree of maximum glacio-isostatic uplift in the Holocene has been considerable: around 300 metres in North America and 307 metres in Fennoscandia, but less in Britain.

The overall look of a landscape owes much to its plate tectonic setting (past and present) and the broad climatic changes that it has experienced. The art of geomorphology is to be able to look at a landscape and start to explain why it looks as it does – and tectonic and climatic factors are usually the first things that geomorphologists invoke. Huge amounts of data have now been collected about the tectonic and climatic histories of many parts of the Earth's surface, all vital for providing ever more sophisticated accounts of landscape development.

Chapter 5
Living landscapes

Most landscapes are covered by some sort of green mantle, whether sparse shrubs in arid areas or dense and tall rainforests within the humid tropics. Indeed, it is rare to find any landscape on Earth that doesn't have some biota. Even the ice-sheet-covered landscapes of Antarctica have been found to contain communities of micro-organisms which play key roles in mineral transformations, as do steep rock walls within high mountain environments. Only a few soils within the most hyper-arid areas of the Atacama Desert in South America have been shown to be biologically dead. Most landscapes are at least partly defined by their vegetation cover as this contributes hugely to the 'look' and 'feel' of the environment, as well as nurturing a whole host of animal life and often supporting human settlement. In many parts of the world, hundreds of years of human settlement has, in turn, brought large changes to the vegetation. Plant communities and the soils they occupy develop in tandem, through the processes of succession recognized by ecologists. In a simple succession on newly exposed land such as a lava flow, a sequence of vegetation communities develop, with hardy pioneer species which can cope with low nutrient levels coming first and altering their own environment in ways that make it more hospitable to other species. Over time, more species colonize and interact with those already present to create a complex community and associated soil.

Climate provides a major control on which plants can grow and which soil types are found, and as climate changes over short, medium, and long timespans, so the plant and soil communities adjust, in turn. Of course, things are rather more complex than this simple model suggests – successions have many other influences upon them. The key point for us, however, is that plant and animal communities involved in succession are not merely a 'living veneer' over the inorganic landscape, but rather they are one important, active component of the landscape palimpsest we identified in Chapter 1. How this living landscape functions is the focus of this chapter. We begin by looking at the ways in which plants and animals are controlled by geomorphology, and then look at the flipside, or how geomorphology is influenced by plants and animals. We end by considering how these two sides of the coin are inter-related, and why this matters for understanding landscapes.

A concern for the interlinkages between geomorphology and the living world is nothing new. Indeed, Charles Darwin carried out some of the most important early work on this topic, such as his investigations on the role of earthworms in creating vegetable mould and aiding denudation. Darwin calculated that:

> there is evidence that on each acre of land, which is sufficiently damp and not too sandy, gravelly or rocky for worms to inhabit, a weight of more than ten tons of earth annually passes through their bodies and is brought to the surface. The result for a country the size of Great Britain, within a period not very long in a geological sense, such as a million years, cannot be insignificant; for the ten tons of earth has to be multiplied first by the above number of years, and then by the number of acres fully stocked with worms; and in England, together with Scotland, the land which is cultivated and is well fitted for these animals, has been estimated at above 32 million acres. The product is 320 million tons of earth.

Natural historians and scientists of the 19th century like Darwin were much more likely to investigate what might now be referred to as 'inter-disciplinary' or 'cross-disciplinary' interactions within the natural world than the generations that followed them – largely because there were few well-established disciplinary boundaries in the early 19th century. They let their interests and observations guide their science, not the other way round. During the 1980s, there was a resurgence of interest in some of the links between plants, animals, and geomorphology, and the names 'biogeomorphology', 'ecogeomorphology', and 'geoecology' have been given to this interface area of science, which has continued to attract interest from geomorphologists and ecologists alike. Links between the organic and inorganic elements of the landscape are not only important to understanding today's landscape, they are also crucial both to interpreting the past and to predicting the future. Following Darwin's early lead, we are also coming to realize that some of the most crucial interactions on the Earth's surface involve extremely small and often overlooked organisms, particularly the vast array of micro-organisms (such as bacteria, fungi, and algae) and the complex, cosmopolitan communities they produce.

How does geomorphology influence plant and animal life, and how does this shape the landscape? At the most obvious level, if we fly over a landscape or observe it from a high point such as a hilltop, it is clear that under natural conditions different vegetation types occupy different parts of the topography. Valley bottoms, for example, support very different soils and plant communities to the upper parts of hillslopes. At a larger scale, as one ascends a mountain, it is clear that the plant communities change with altitude. Not only topography but also underlying geology can be highly important. Moving from basalt to limestone terrain, for example, one encounters differences in relief, soils, and plant communities, all associated with the fact that different rock types weather and erode in different ways as a result of their varying resilience and mineral composition. Thus, ecologists have recognized that, whilst climate provides the crucial global-scale control on biodiversity, geology and topography provide important regional-scale influences on the numbers and types of plant and animal species. Much recent work has gone into trying to provide some quantitative measure of topographical complexity and link this to biodiversity at different scales.

Geomorphology also influences plant and animal communities in more active ways. Many geomorphological processes involve the movement of sediment and/or the creation or exposure of new rock surfaces. For example, in arid areas with a suitable supply of sand-sized sediment, winds erode, transport, and deposit sand, creating often vast, mobile dune systems. Similarly, within alluvial river systems, fluvial processes erode, transport, and deposit sediment to produce bars, islands, and overwash deposits. Mass movements, such as rockfalls, create newly exposed rock surfaces, as does coastal erosion of rocky coastal platforms. So, what we can see is that through the action of geomorphic processes, new components of the landscape are constantly being produced and altered. These dynamic surfaces become colonized by plants and animals, creating more opportunities for biodiversity. In rivers, for example, sedimentary deposits not only provide a surface for colonization by riparian (river bank) vegetation, but also habitats for a range of fish species. Generally speaking, these dynamic landform influences on biological communities act at a relatively small scale, but in some instances, such as large lava flows, they can alter very large areas in a very short space of time. We can interpret these influences as an example of 'geomorphological engineering', or the action of geomorphological processes to enhance the biodiversity of an area through introducing dynamism and creating new habitats. The result of both this geomorphological engineering and the influence of topography on communities and biodiversity reviewed above is to add complexity to the landscape. Some scientists even go so far as to say that the links between biology and geomorphology produce an evolutionary geomorphology, one which is being altered as communities evolve and change.

Plants and animals are not merely passive occupiers of the Earth's surface; they play a key role in many geomorphological processes and can create unique landforms. Both these effects are sometimes called 'ecological engineering' by ecologists. What this phrase means is that the plants, animals, or micro-organisms involved engineer the environment to make it more favourable as a habitat – both for themselves and for other species. Both these effects, and those of the parallel 'geomorphological engineering' we identified above, enhance local biodiversity and productivity.

Let's focus first on the role of plants, animals, and micro-organisms in creating unique landforms, as this is perhaps the clearest manifestation of ecological engineering. Beaver dams, for example, can cause large-scale alterations to the landscape, as shown from parts of the USA where large areas of forested land have become

9. Part of a beaver dam, Ontario, Canada

inundated by ponds, and 'beaver meadows' dominated by thick willow growth have developed (Figure 9). The beaver dams are constructed from trees felled by beavers, and together the dams and ponds alter the surrounding fluvial and riparian habitats, facilitating some other species, such as water-loving birds and small mammals. There are many other examples of constructions actively made by organisms as structures to aid catching prey or for courtship which are also landforms, but these are often very small and transient. Many animals burrow and create mounds of sediment as dwellings, which also leave a clear imprint on the landscape – such as badger setts, pocket gopher mounds, and termite mounds. Termites are amazingly effective agents of ecological and geomorphological engineering – creating huge, complex, and resilient edifices as well as moving vast amounts of sediment around the landscape. Other constructions, such as coral reefs, tufa deposits, and stromatolites, involve a combination of dead and alive animal, plant, and micro-organic matter. In these instances, the constructions are partly made from skeletal material, and partly from debris (organic and inorganic) cemented with chemical and biochemical precipitates, and the live organisms live on and within the dead ones.

One landform created almost entirely from dead material is the log jam, in which rivers become completely dammed or partially blocked by logs (better known as 'large woody debris', or LWD). Dead wood falls into streams and rivers as a result of death or disease, or perhaps mass movements, fluvial erosion of riparian trees, or even animal attack. This dead wood accumulates, and then other sediment and debris builds up behind it. One extreme example of this process occurred on the northern Red River in Louisiana, before European settlement. Over at least a 200-year period, a series of log jams affected a 260-kilometre-long section of the river, with the jams accumulating into a huge log raft and producing a whole series of lakes. The log raft was gradually removed during the 19th century by river steamers with heavy cranes and the lakes were drained.

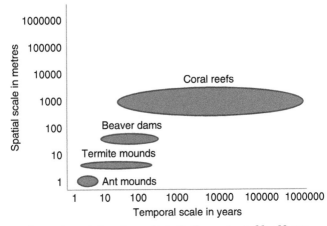

10. Longevity and size of some biologically constructed landforms

As can be seen in Figure 10, biologically constructed landforms range widely in scale and longevity, but in comparison with the whole spectrum of landforms illustrated in Figure 2, they are relatively small, and both form and decay quite quickly. Exceptions include large and long-lived coral reef constructions which can persist within a landscape for millions of years and, in the case of the larger ones, such as the Great Barrier Reef, are several thousand kilometres in dimension. One key question here, given the range of biologically produced landforms, is: would the world's landscapes look different if there were no plants, animals, or micro-organisms? Clearly, they would look superficially different, in that there would be no green mantle, but would the basic lineaments of the landscape be different? This question has been the source of much geomorphological debate in recent years, but the general agreement is that at the large scale, it wouldn't look much different. This is because of the overwhelming importance of tectonics and climate as the basic controls on landscapes at this scale. However, at the more local scale, there are many instances of landscapes whose very fabric and appearance have been heavily influenced by plant and animal action. This question is not purely of relevance to today's landscape, as over the long history of the Earth, life has developed through evolution alongside changing geomorphological landscapes, and early plant and animal communities may have imprinted themselves rather differently on the landscape. Such palaeo-landscapes are occasionally visible in geological sections.

Perhaps the more crucial role that plants, animals, and micro-organisms play within the landscape is not directly through creating landforms, but indirectly through their influence on Earth surface processes. There is a bewildering array of biological influences on geomorphic processes, but they largely fall into two types, either accelerating or retarding the rates of process operation. There are many instances where animals accelerate Earth surface processes, as seen, for example, in the way in which grazing of cattle and other large animals can lead to accelerated soil erosion, through removing the protective layer of vegetation. Cattle have also been observed to accelerate the erosion of river banks as they come down in herds to drink. Smaller animals, such as ants, termites, earthworms, and ground-dwelling birds, have huge effects on soil cycling and mineral transformations, through activities which move and alter sediment (for example, through their guts). Micro-organisms are similarly highly effective agents of mineral transformation, and there is much evidence to suggest that they can radically enhance weathering rates. Some species of lichen, for example, are able to etch their way into a rock surface, producing small-scale pits as a result (see Figure 11). At a larger scale, vegetation communities as a whole have been shown to dramatically enhance the rates of rock weathering. If you compare the rate of weathering under soil and vegetation communities (especially forested ecosystems) with that on bare rock surface, you will find that they are several orders of magnitude higher. This largely results from the production of organic acids and carbon dioxide, which are highly effective agents of chemical weathering.

11. 'Rock-eating' lichens, Golden Gate Highlands National Park, South Africa

Whilst there are many examples of biological enhancement of denudation, there are also examples of biological enhancements of other processes, notably deposition. Thus, vegetation can also be observed to accelerate the deposition of sediment on dunes, salt marshes, and mangrove swamps. Dune vegetation, for example, causes alterations in air flow within the boundary layer (the zone immediately above the ground surface), encouraging local reduction in wind velocities and the deposition of entrained sand. Some types of vegetation do this more effectively than others, and their accretionary roles can also be seen as an example of ecological engineering, as it makes the local environment even more stable and favourable for colonization by other species.

Organisms can also retard the rate of Earth surface processes, with vegetation producing some of the most spectacular effects. Forested slopes, for example, generally experience much lower rates of erosion than similar slopes without tree cover. The trees intercept rainfall, reduce local windspeeds, and physically bind together the soil. These effects together reduce the rate of water- and wind-borne erosion and the frequency of mass movement activity. Even where plants do not enhance deposition of sediment, their role in protecting surfaces against erosion can be extremely important. The combined effects of reducing erosion and enhancing accretion can be termed 'bioprotection', as they increase the resilience of surfaces in the face of denudation. Bioprotection is often used as a form of environmental management through planting vegetation to reduce erosion risk and encourage deposition, so as to conserve landscapes such as coastal dunes. However, it is important to remember that under natural conditions these bioprotective roles go hand in hand with biological enhancement of denudation and landform construction, and so humans' attempts to prioritize one may not be successful or entirely desirable. We have also to ask whether biological influences on Earth surface processes are, overall, significant. To put it another way, would the Earth's landscapes function differently if there were no plants, animals, and micro-organisms? Opinions vary on this, but there is a large body of evidence which suggests that soils would be hugely different without biologically induced transformations of rocks and sediments, and that denudation would be quicker in an abiotic world. However, it is extremely difficult to answer this question, as there is little quantitative and comparable data on process rates with and without organic influences, and thus little hard evidence on which to base one's answer.

Despite the insights of the early landscape detectives, it has taken geomorphologists some time to acknowledge the role of organisms in shaping the landscape we see around us. Largely, this was because of the difficulty of quantifying plant, animal, and micro-organism effects and including them within our geomorphological models. Some extremely patient and detailed monitoring of, for example, the amount of sediment moved by ant communities on Australian hillslopes, has recently been undertaken, and we now have a better (but by no means really adequate) grasp of how plants, animals, and micro-organisms interact with geomorphology. In parallel, ecologists have been similarly slow to get to grips with how geomorphology (in the form of both landforms and processes) influences plant and animal communities, with issues of the difficulty of quantification proving similarly hard to tackle. We are now poised at an exciting moment in time when we can start to consider in detail the two-way interactions between geomorphology and ecology as they shape landscapes. This matters because it provides the key to providing better management solutions for local and global environmental problems, such as erosion, biodiversity loss, and global change, as we shall see in more detail in Chapters 6 and 7.

Chapter 6
Landscapes and us

Paul Crutzen, who won the Nobel Prize for Chemistry in 1995, coined the term 'anthropocene' in 2000 to cover the last 300 or so years of geological time – from the start of the Industrial Revolution. His point in doing this was to illustrate the important role that humans are now having, for the first time, in influencing the biosphere, atmosphere, and oceans. Whilst the 'anthropocene' concept has proved to be controversial, it provides a useful starting point for this chapter, as within it are subsumed many issues relating to humans and landscapes. The 'human impact' on many components of the Earth system is now well documented, and in turn there have been many accounts of the way in which events and processes within the Earth's system impinge upon us. What concerns us in this chapter is an assessment of how humans have affected, and continue to affect, the landscape through their impact on geomorphological processes and landforms, and how the landscape affects us through geomorphological hazards. Both are key symptoms, often inter-related, of life in the anthropocene. We conclude the chapter with a discussion of how geomorphological skills can be applied to remediating human impacts and helping us cope better with hazards.

The study of human interrelations with geomorphology can be called 'anthropogeomorphology'. This rather unwieldy term occurs especially in German-language literature, and can be taken to mean the study of the human role in creating landforms and modifying the operation of geomorphological processes such as weathering, erosion, transport, and deposition. However, it can be taken rather more broadly to mean the study of all geomorphologically based interactions between humans and landscape. Geomorphologists largely ignored human dimensions of the landscape until the late 20th century, perhaps because of the difficulty of quantifying them, or because of the assumption that natural forces were dominant and humans mere passive inhabitants of the landscape.

Let's look firstly at the direct and indirect human impacts on landscapes, as summarized in Table 1. Some landscape features are clearly produced by direct anthropogenic processes, with often regular and unnatural-looking shapes, and are frequently created deliberately and knowingly. The key direct anthropogeomorphic processes are construction, excavation, and hydrological interference. Typical landforms created in this way are sea walls, embankments, spoil heaps, agricultural terraces, mines and quarries, canals and reservoirs. In 1994, Robert B. le Hooke attempted to quantify the amount of earth moved in such ways by human activity in comparison with natural denudation. He suggested that in the USA alone, excavation for housing and other construction moved 0.8 billion metric tons of sediment per year, mining 3.8 billion, and road construction 3 billion, giving a total of 7.6 billion metric tons per year in the USA. Looking globally, he estimated some 30 billion metric tons per year of sediment is moved in this way. In comparison, over the whole planet, rivers transport an estimated 14 billion metric tons of sediment to the oceans per year, whilst an additional 40 billion tons is moved short distances within river basins, glacial action moves 4.3 billion, and wind a further 1 billion tons. Thus, it is fair to say that humans now move a comparable amount of sediment per year to natural denudational processes.

Human activity doesn't stop there. Indeed, the major contribution of humans to anthropogeomorphology is through indirect

Table 1. Major anthropogeomorphic processes

Direct or indirect	Nature of process	Examples
Direct	Constructional	Tipping, moulding, ploughing, terracing, reclamation
	Excavational	Digging, cutting, mining, blasting, craters, trampling
	Hydrological interference	Flooding, damming, canal construction, dredging, channel modification, draining, coastal protections
Indirect	Acceleration of erosion and sedimentation	Agricultural activity and vegetation clearance, engineering (especially road construction and urbanization), incidental modifications of hydrological regime
	Subsidence (collapse, settling)	Mining, groundwater and hydrocarbon removal, thermokarst (permafrost melting)
	Slope failure (landslides, flows, accelerated creep)	Loading, undercutting, shaking, lubrication
	Earthquake generation	Loading (reservoirs), fault plain lubrication
	Weathering	Acidification of precipitation, accelerated salinization, laterization

acceleration of natural processes, which is much harder to quantify, but certainly exceeds that of direct processes. A wide range of human activities, from farming to fishing to heavy industry, can accelerate erosion and deposition of sediment. For example, cutting down forests and replacing them with grazing land (one example of what is often, rather blandly, called 'land cover change') can result in huge amounts of sediment being eroded from one area, transported by rivers, and deposited elsewhere (Figure 12). Similarly, mining and quarrying, as well as the direct effects on the landscape noted above, can cause subsidence and collapse of undermined terrain, as can the extraction of hydrocarbons and groundwater from sensitive deltas. Another huge impact of humans on the landscape relates to their acceleration or encouragement of landslides and other mass movements. Building on slopes, clearing vegetation away from slopes, and altering the drainage on slopes can all help trigger mass movements such as landslides. At a more local scale, human impacts are also very important in accelerating weathering through the production of acidified rain. As we shall see in the next chapter, human-induced global warming is already resulting in

12. Dongas, or eroded valley sides, Swaziland

additional indirect impacts on the landscape and will continue to do so for many centuries.

Finally, there are situations where direct human interventions in the landscape can have linked unforeseen and unwanted indirect impacts. For example, there are many records of attempts to reduce coastal erosion which, far from solving it, only exacerbated it. Examples include the role of sea walls in causing beach scour. In the main, these problems have arisen because of a lack of understanding of the overall geomorphological system in which one landscape component is situated. Thus, protecting one bit of coast which forms part of a natural sediment circulation system without realizing its larger setting will, of course, have unanticipated consequences elsewhere within the system.

Can we disentangle human and natural changes to the landscape? Often, because of the complexity of direct and indirect human impacts, and the fact that many work with natural processes, this is hugely difficult. Sometimes, however, the timing or nature of the change can help detect what is responsible. A good example of the problems of detecting human impacts is that of the development of arroyos. The name 'arroyo' comes from the Spanish term for watercourse, and is used to describe gulleys cut into the bottom of valleys in the south-western USA. Intriguingly, large numbers of arroyos developed in this area between 1865 and 1915. A range of human activities have been cited as the cause of this erosion spasm, such as timber felling, overgrazing, harvesting of grass, compaction along tracks, channelling of runoff from trails and railways, and disruption of valley-bottom sods by livestock. Natural environmental changes could also be responsible, as similar periods of arroyo incision have been found to have occurred repeatedly before the arrival of Europeans. For example, periods of aridity can trigger arroyo development, as they reduce vegetation cover, exposing the ground to erosion, whilst periods of increased storminess can enhance erosive runoff. So, how can geomorphologists aid our understanding of such human impacts and prove whether or not human actions have been responsible for gullying or arroyo formation? One of the main tools required is the ability to date periods of enhanced gully erosion or arroyo incision, as often the key to answering a question about 'what is responsible for a change?' is knowing when exactly that change occurred. The burgeoning number of new radiometric dating techniques discussed in Chapter 3, in tandem with well-established, but continually improving, techniques such as dendrochronology (the use of tree-ring evidence to date events in the landscape), can help geomorphologists to determine when pulses of erosion occurred. Cross-checking can then be done of these timings with any severe climatological events and with histories of human habitation and agriculture. However, often both climatological and human events occur around the same time, and so other tools then have to be used to deduce which was responsible, or whether it was indeed the combination of human and natural processes which caused the arroyos.

Accelerated erosion in the anthropocene

Erosion is a process that has operated constantly since the Earth was created. Erosion of soil by water has likewise been an entirely natural phenomenon ever since soils appeared. However, since humans first inhabited the Earth, their activities have caused spasms of accelerated soil erosion associated with land cover and land use changes. In Classical times, it was noted that slopes in areas such as Greece, Turkey, and the Levant had been destabilized by deforestation and over-grazing. Undoubtedly, such actions as the deliberate setting of fire, the adoption of pastoralism and agriculture, deforestation, urbanization, and the use of machinery to move and disturb the soil, have all contributed to accelerating rates. Bill Ruddiman hypothesizes that the anthropocene actually started some 8,000 years ago, much earlier than proposed by Paul Crutzen, as a response to the beginning of agricultural civilizations and their impacts on the environment.

Deforestation has been a crucial cause of accelerated soil erosion in many areas. Forests protect the underlying soil from the direct effects of rainfall, generating an environment in which erosion rates tend to be low. The canopy shortens the fall of raindrops, decreases their velocity, and thus reduces their kinetic energy. Most canopies reduce the erosion effects of rainfall. The presence of humus in forest soils absorbs the impact of raindrops and gives them extremely high permeability. Thus forest soils have high infiltration capacities. Forest soils also transmit large quantities of water through their fabrics because they have many macro-pores produced by roots and their rich soil fauna. They are also well aggregated, making them resistant to both wetting and water drop impact. This superior aggregation is a result of the presence of considerable organic material, which is an important cementing agent in the formation of large water-stable aggregates. Furthermore, earthworms also help to produce large aggregates. It is therefore to be expected that with forest removal, rates of soil loss will rise and mass movements will increase in magnitude and frequency. The rates of erosion will be high if the ground is left bare; under crops, the increase will be less marked. Furthermore, the method of ploughing, the time of planting, the nature of the crop, and the size of the fields will all influence the severity of erosion.

A good example of using long-term sedimentation rates to infer long-term erosion rates is provided by a study of the Kuk Swamp in Papua New Guinea. This identified low rates of erosion until 9,000 BP (before present), when, with the onset of the first phase of forest clearance, they increased from 0.15 cm 1000 y^{-1} to about 1.2 cm 1000 y^{-1}. Rates remained relatively stable until the last few decades when, following European contact, the extension of anthropogenic grasslands, subsistence gardens, and coffee plantations produced a rate that was very markedly higher: 34 cm 1000 y^{-1}.

There is increasing evidence to suggest that silty valley fills in Germany, France, and Britain, many of them dating back to the Bronze Age and the Iron Age, are the result of accelerated slope erosion produced by the activities of early farmers. During the Holocene in Britain, among the formative events that have been identified are initial land clearance by Mesolithic and Neolithic peoples; agricultural intensification and sedentarization in the late Bronze Age; the widespread adoption of the iron plough in the early Iron Age; settlement by the Vikings; and the introduction of sheep farming.

There is some evidence that soil erosion is becoming a more serious problem in parts of Britain, in spite of the fact that the country's rainfall is much less intense, and so less erosive, than in many parts of the world, and despite the fact that the causes of accelerated soil erosion are well documented and understood. The following practices may have caused this state of affairs:

- Ploughing on steep slopes that were formerly under grass, in order to increase the area of arable cultivation.
- Use of larger and heavier agricultural machinery, which tends to increase soil compaction.
- Use of more powerful machinery which permits cultivation in the direction of maximum slope rather than along contours. Rills often develop along the wheel ruts ('wheelings') left by tractors and farm implements, and along drill lines.
- Use of powered harrows in seedbed preparation and the rolling of fields after drilling.
- Removal of hedgerows and the associated increase in field size. Larger fields cause an increase in slope length and thus a higher risk of erosion.
- Declining levels of organic matter resulting from intensive cultivation and reliance on chemical fertilizers, which in turn lead to reduced aggregate stability.

- Widespread introduction of autumn-sown cereals to replace spring-sown cereals. Because of their longer growing season, autumn-sown cereals produce greater yields and are therefore more profitable. The change means that seedbeds with fine tilth and little vegetation cover are exposed throughout the period of winter rainfall.

There is likewise evidence from other areas of the world, such as sub-Saharan Africa, for an upward trend in soil erosion linked to changes in agricultural practice. However, there is also a large body of evidence which reveals that increases in population do not necessarily lead to soil erosion within African countries. Extensive studies from the Machakos district in Kenya have revealed that 'more people, less erosion' can occur. So, human impacts do not always accelerate geomorphological processes and we can, and should, learn from our mistakes.

The other side of the anthropogeomorphological coin is the role that geomorphological hazards play in affecting humans. Geomorphological hazards can be defined as any Earth surface process which can cause loss of life and/or economic disruption in the form of loss of property and/or livelihoods. Obvious examples are abrupt and large-scale mass movements which can destroy large areas of habitation, and the slower, chronic process of soil erosion which can lead to loss of topsoil and agricultural collapse. Some different types of hazard are listed in Table 2. We can thus make a useful distinction between catastrophic and pervasive hazards. Geomorphological hazards are a subset of natural hazards which cover any event or process within the Earth system that affects people. The major natural hazards in terms of both loss of life and damage to property are tectonic and climatic (such as earthquakes and hurricanes). Both tectonic and climatic hazards can involve secondary geomorphological hazards. For example, many earthquakes trigger off landslides which themselves can cause more damage than the earthquake itself, whilst hurricanes can be accompanied by storm surges that can create massive

Table 2. Examples of geomorphological hazards

Arid zones	Coastal
Dune encroachment and sand drift	Sea level change
Soil deflation	Dune blowouts and encroachment
Arroyo formation	Cliff retreat
Dust storms	Salt marsh siltation
Fan entrenchment	Coastal progradation
Flash floods	Spit growth and breaching
Salt weathering	Storm surges
Ground subsidence	Tsunami rip currents
Tundra areas	**General**
Thermokarst formation	Mass movement (mudflows, quickclays, debris flows, rockfalls, landslides, etc.)
Frost heave	Karstic collapse
Thaw floods and ice jams	River floods
Glacier surges and glacier dams	Shifting river courses
Avalanches	Lake sedimentation

Tundra areas	General
Jökulhaups (outwash floods)	Soil erosion
	River bank erosion
	Tectonic and seismic activity. Volcanic activity (explosions, lava flows, tephra fallout, ballistic projectiles, lahars, toxic gas, pyroclastic density currents). Ground subsidence

amounts of coastal erosion, again often causing considerable damage to property.

Although high-magnitude, low-frequency catastrophic events, such as hurricanes or earthquakes with their concomitant geomorphic hazards, gain attention because of the immediacy of large numbers of casualties and great financial losses, there are many more pervasive geomorphological changes which are also of great significance for human welfare and livelihoods. These may have a slower speed of onset, a longer duration, a wider spatial extent, and a greater frequency of occurrence. Examples include weathering phenomena and soil erosion.

Indeed, there is a great diversity of geomorphological hazards, many of which occur in particular environments whilst others are worldwide in occurrence. One major worldwide category is mass movements, such as rockfalls, debris flows, landslides, and avalanches. There are also various fluvial hazards, such as floods and river channel changes. In volcanic areas, there are disasters caused by eruptions, lava flows, ash falls, and lahars (volcanic debris flows). Seismic activity is another type of hazard associated with tectonic activity. In coastal environments, inundation and erosion caused by storm surges, rapid coastal erosion and siltation, sand and dune encroachment, shoreline retreat, and sea level rise are all important hazards. In glacial areas, hazards may be posed by such phenomena as glacial surges, outwash floods, and damming of drainage. Permafrost regions may be hazardous because of ground heave, thermokarst development, icings, and other such phenomena. There is also a wide range of subsidence hazards caused by solution of limestone, dolomites and evaporites such as gypsum or halite, degradation of organic soils, hydrocompaction of sediments, and anthropogenic removal of groundwater and hydrocarbons. In desert regions, hazards are posed by wind erosion and deflation of susceptible surfaces, dust storm generation, and by dune migration. More generally, water erosion causes soil loss and gully or badland formation, while weathering can be a threat to a wide range of engineering structures.

How are the two sides of anthropogeomorphology related? Does human impact on the environment influence the occurrence of hazards? And do dramatic natural events and processes make a difference to the human impacts on geomorphology? Much recent research indicates that geomorphic hazards can be increased or triggered by human activities, and in particular by land use and land cover changes. These are also the key ways in which humans affect geomorphology, as we have seen. In turn, it is clear that major hazardous events such as landslides triggered by earthquakes can produce major changes to the landscape which in turn may cause a flurry of anthropogenic sediment movement as rebuilding occurs. Set within the context of climate and tectonics providing the major controls on landscape, human impacts might seem quite puny, but they can be critical in altering natural process regimes, and this in turn can often trigger or enhance hazards. Humans are thus caught in a vicious cycle in many parts of the world, whilst in other areas our footprint on the landscape is much less marked or natural hazards much less common.

How can geomorphologists contribute to reducing geomorphic hazards and mitigating human impacts on the landscape? This is one of the key challenges facing what is often called 'applied geomorphology'. Geomorphologists have applied their skills to problems facing humanity for many years. The great American geomorphologists of the second half of the 19th century – G. K. Powell, C. Dutton, W. J. McGee, and G. K. Gilbert – were employed by the US Department of the Interior to undertake surveys to enable the development of the West, and R. E. Horton,

Anthropogeomorphology of arid environments

Arid environments illustrate the diversity and severity of geomorphological hazards, as well as a burgeoning human impact on the environment. In recent years, as urbanization has spread within deserts and other forms of development have taken place, the importance of geomorphological hazards has become increasingly apparent. Wind erosion and dust storms, sand dune encroachment, water erosion, slope instability, subsidence, lake level changes, and salinization are all important hazards within the arid realm. In many arid areas, rapid urbanization and development are also associated with some of the most dramatic direct and indirect human impacts on sediment movement yet seen on Earth.

Arid environments are often highly dynamic geomorphologically, primarily as a result of their variable climatic conditions. Where human settlements are found, such dynamism can translate into hazards, whilst, in turn, human activities can make these hazards worse. A run of dry years is often ended by dramatically wet conditions, or as the climatologists would put it, arid environments suffer from high inter-annual variability in precipitation. As a consequence of this climatic variability, plant and animal communities are also dynamic and change over time. Varying amounts of moisture and associated changes to vegetation lead to changing amounts of aeolian and water-borne erosion. Dunes can be stable for many years as a result of vegetation preventing aeolian sand transport, but a few dry years can upset this balance and cause rapid erosion and reactivation of dunes. Such dune movement can pose a hazard to human settlements and agriculture in the area. However, often, dune movements can be triggered by human activities – as humans interfere with vegetation through stock grazing, for example. Much effort has been spent in recent years trying to understand, and thus be able to control, the dynamic arid environments, but there remains an essential conflict between the expectation of stable, economically developed conditions and the inherent dynamism and variability of the desert landscape.

A highly dramatic example of human impacts on the arid landscape comes from the United Arab Emirates (Figure 13), where the volume of direct sediment movement must be extremely high over recent years, with the building of offshore island complexes such as Palm Island and The World. Vast amounts of sand have been quarried from inland and deposited in the shallow waters of the Arabian Gulf. Huge road-building projects are being undertaken, as well as rapid urban expansion and the building of myriad projects such as golf courses and leisure complexes. Past experience would lead us to suspect that such intensive anthropogenic activity will have an enhancing effect on geomorphic hazards.

one of the founders of modern quantitative geomorphology, was active in the soil conservation movement generated by the 'Dust Bowl' conditions of the 1930s. A great deal of significant work was undertaken in the USA under the direction of H. H. 'Big Hugh' Bennett on soil erosion and runoff processes, while workers like W. S. Chepil did fundamental research on the causes and controls of wind erosion on agricultural land. During the Second World War, geomorphologists like Ralph Bagnold were involved with

13. Bulldozing the desert – using inland sand deposits in the United Arab Emirates for offshore building projects

trafficability studies in the Western Desert and elsewhere, and coastal geomorphologists were concerned with establishing the suitability of beaches for amphibious operations.

Applied geomorphology is a popular pursuit today, for a range of reasons. The range of techniques available to geomorphologists can now provide useful answers to many questions, there is a lot of funding now available for applied projects, and an ever increasing number of problems seem to need solving as the world's population grows and development proceeds apace. So what are the essential geomorphological skills which can be used in applied studies? One of the key skills, and often greatly underestimated, is an 'eye for country', or the ability to interpret landscapes. Geomorphological mapping skills are also highly valuable, as they enable maps to be produced of the distribution of hazard-prone areas. The geomorphological understanding of landscape as a system is also highly valuable, as it highlights links between different processes and landforms which might not be obvious to other experts. Such knowledge can be very important in preventing unintended consequences of environmental modifications. The geomorphological ability to see things within a long-term context is also highly valuable in attempts to understand current hazards and human impacts. As an example, we can identify a number of different contributions geomorphologists can make to understanding and reducing the impact of geomorphic hazards. These include: the mapping of hazard-prone areas; constructing the history of occurrence of past hazardous events; establishing their frequency and magnitude; predicting the occurrence and location of future events; monitoring geomorphological change; and using knowledge of the dynamics of geomorphological processes to advise on appropriate mitigation strategies. Of course, as most geomorphological hazards are highly related to climatic and tectonic events and processes, applied geomorphologists commonly work closely in association with climatologists and Earth scientists. One of the areas of applied geomorphology that should develop rapidly over the coming decades is an involvement in geoengineering projects, or schemes which use the natural environment to help in management and mitigation of future climate changes. Could, for example, any parts of the landscape, such as karst cave systems, be used to store carbon? Perhaps the most important aspect of applied geomorphological research in relation to future climate change is the ability to predict the future, as we shall see more fully in the next chapter.

Chapter 7
Landscapes of the future

The biggest challenge facing landscapes in the future is commonly
seen to be climate change. Climatic change as a result of global
warming, allied with the growth of local human impacts on the
environment around the globe, will have major effects on our
landscapes. These effects will not be felt on plants and animals
alone, but will be mediated by, and mixed up with, impacts on
the geomorphology. Geomorphologists can and should contribute
to predicting landscape futures because of these links between
climate, ecology, and geomorphology, but their contribution to
global change science has, so far, been undervalued. The
Intergovernmental Panel on Climate Change (IPCC) and other
organizations have produced major syntheses of information about
the current state of the environment and how it is likely to change
in future as a result of climate change. However, geomorphology
has played a surprisingly small part in such exercises, which tend
to have been dominated by climatologists, oceanographers, and
ecologists. But, as we have seen in Chapter 1, the foundation of
the landscape is geomorphology and any changes in climate and
ecology will have knock-on impacts on geomorphology, which,
in turn, will condition and influence the climate and ecology.
In this chapter, we aim to bring landscape and geomorphology
fully into the global change arena, by considering, on the one hand,
what impacts climatic change will have on landscapes and, on
the other hand, how such landscape changes will affect climate

change. Underlying our discussion is the important point that climate change is only one driver of landscape change, and we need to weigh up the relative impacts of global climate change and local human impacts on landscapes. So, the challenge to future landscapes might be more complex than commonly thought. The key concept to grapple with here is that of 'feedbacks', which we will explain first.

Looking at landscapes as a series of interconnecting systems illustrates the concept of feedbacks. Feedbacks can be positive (reinforcing) or negative (dampening). A positive feedback occurs when a change in a system becomes magnified and enhanced – if, for example, a person starts to eat larger meals, they get fatter and often then feel the need to eat more, becoming fatter, and so on. A negative feedback occurs when a change to a system becomes reduced or dampened over time. For example, if a slope becomes unstable and fails through mass movement processes, it then becomes stable and further movements occur less frequently.

Looking at how climate change will affect landscapes, it is predicted that some landscapes will be particularly prone to change. Why are some landscapes more likely to be 'geomorphological hotspots' than others? First, some landscapes are close to important thresholds. Some landforms and Earth surface processes are prone to change across crucial thresholds of temperature and precipitation. For example, the melting of ice is strongly temperature-dependent and, for example, permafrost can only exist where mean annual temperatures are negative. Once mean annual temperatures exceed $0°C$, permafrost becomes unstable and decays. Thus, as global temperatures rise, the distribution of permafrost will shrink polewards and with altitude, and the depth to which the ground thaws in summer will increase across many areas. Similarly, the state of glaciers is largely controlled by the relative significance of melting and of snow nourishment (or what glaciologists call the balance between ablation and accumulation) and these in turn depend on both temperature and the amount of precipitation. Some glaciers which experience relatively warm conditions with low snowfall today could thus experience very rapid retreat in future if temperatures rise but snowfall doesn't increase – as the balance between ablation and accumulation will tip heavily towards ablation.

Some landscapes are likely to be particularly prone to alteration as a result of climate change because of the 'compound effect', that is, other influences such as local human impacts which may work synergistically to produce landscape change. The US Dust Bowl of the 1930s, for example, was caused not only by a run of dry, hot years, but also coincided with a phase of land use intensification involving ploughing up of prairie grasslands. Indeed, land degradation on desert margins (desertification) is a phenomenon that is often at its most intense when climatic pressures (such as drought) and human pressures (such as over-cultivation) coincide. We might suggest that landscapes already subjected to many local human pressures may be particularly sensitive in the face of future climatic change. However, it is also clear that some landforms are robust, while others are not. What we mean by 'robust' is that they are not highly sensitive to external factors changing. Robustness is often a function of the nature of the materials making up the landscape. Muddy and sandy coastlines, for example, are more prone to erosion than hard-rock coastlines simply because the materials that they are composed of are more easily eroded by wave and tidal action.

Some landscapes will be prone to climate change effects simply because the climate is predicted to change more in some areas. For example, the degree of temperature increase is predicted to be particularly great in high northern latitudes (for example in northern Canada), so that ice in such regions may melt rapidly. Warming will have an especially strong impact on river behaviour in areas where winter precipitation currently falls as snow. This is because snow persists on the landscape until the start of the melt season when it contributes to a sudden meltwater pulse within the river systems. If climate change leads to winter precipitation in these regions falling as rain, there will be major effects on annual river regimes.

Let's look now at some specific examples of geomorphological hotspots and how climatic change in the future will dramatically affect sensitive coastal, desert, and polar landscapes.

It is difficult to measure the extent of the world's coastline. However, even though the global coast covers a relatively small proportion of the total surface of the Earth, it is an extremely important area. Recent studies indicate that around 1.2 billion people live within both 100 metres of sea level and 100 kilometres of the coast (that is, on relatively low-lying coasts). At 1990 figures, this is around 23% of the total global population. The average population density of these coastal areas is 112 people per square kilometre which is almost three times the average population density of 44 people per square kilometre. Most low-lying coasts can be seen to be geomorphological hotspots in the face of future climatic change, especially those which contain coastal dunes and beaches, mudflats, marshes and mangrove swamps, and coral reefs. However, not every example of these types of coast will be equally threatened by climate change, and the impacts of climate change will often be highly complex. In essence, climate change will affect all these sensitive coasts through changes in sea level and alterations in storminess, whilst coral reefs will further be affected by alterations to sea surface temperatures and ocean chemistry.

Coastal landscapes are regulated by sea level. The height of the sea relative to the land controls the location of coastal erosion and deposition, as well as influencing the distribution of plant and animal communities. If global temperatures climb, then so will sea levels. This is partly because ocean waters, like mercury in a thermometer, will increase in volume as they get warmer. This is termed the 'steric effect'. Sea levels will also rise because of melting of various types of ice, especially glaciers and ice caps.

In the past, there has been a considerable diversity of views about how much sea level rise is likely to occur by 2100. In general, however, best estimates have now settled at just under 50 centimetres by 2100. This implies rates of sea level rise of around 5 millimetres per year, which compares with a rate of about 1.5 to 2.0 millimetres during the 20th century. However, should Greenland ice melt at a faster rate than is currently predicted, then this amount of rise will be exceeded. Some areas of coast will suffer much higher rates of sea level rise than this, however, especially areas with local subsidence problems. Subsidence can result from crustal (tectonic) movements, isostatic adjustments caused by the addition or removal of mass from the Earth's crust, and extraction of solids and liquids (as under some mega-cities like Tokyo, Bangkok, and Los Angeles). For example, the extraction of oil and groundwater from many deltaic coastlines has led to subsidence as the voids left have collapsed or become compressed by the overlying sediment. Particularly fast rates of subsidence have been noted on the Mississippi delta, as well as across much of south-east England.

What will happen to sensitive coastlines facing rapid rates of sea level rise? Dire predictions of extensive areas of coastal inundation and loss of land have been made in the past. More recent predictions are that whilst erosion will increase and flood risk will also be enhanced, many coastlines will react dynamically to sea level rise, with readjustments of the sediment systems to the new conditions. But, of course, these readjustments can only occur where humans have not completely altered the natural landscape, through enterprises such as hard engineering of coastal protection schemes or building cities such as Venice and New Orleans. Sea level rise is also likely to have complex effects because coastal landscapes do not behave as simple, linear systems and sea level rise is not the only control on them. Delta coasts provide a good example of coastal complexity and dynamism. Deltas form where large volumes of fluvial sediment accumulate on shallow continental shelves. Many of the world's great deltas started to form around 6,000 years ago as the rapid rate of sea level rise experienced at the start of the Holocene began to decrease. The accumulation of deltas reflects the changing balance over space and time between the delivery of fluvial sediment and the erosive action of waves and tides. For example, the Mississippi delta has a complex history of depositing fluvial sediment at the coast within different 'lobes'. Every few centuries, the delta system switches to deposit sediment in a new lobe, and the older one becomes abandoned by the present channel system and starts to become eroded by the sea. Human action has further complicated the picture, by interrupting river flows with dams and controlling them with levées. Thus, for many years sediment has not moved naturally across the Mississippi delta. What this means is that some parts of the Mississippi delta coast are now starved of sediment and are easily eroded. And, as we have seen above, humans have also enhanced subsidence across the delta, leading to locally enhanced rates of sea level rise. Events such as Hurricane Katrina throw further complexity into the picture, causing episodically high rates of erosion in some areas, but also bringing deposition of sediment to other parts of the delta. Recent studies show that 24 out of 33 major world deltas are subsiding because of a range of human activities, with 85% having experienced major flooding in recent years.

As sea level rises, coastal erosion of deltas will increase, releasing some sediment into the ocean. Some of this sediment may be lost to sea, whilst some may become deposited elsewhere within the deltaic system. However, another consequence of sea level rise will be the elevation of the base level of the river system – thus decreasing the energy in the lower reaches of the river system and encouraging deposition of sediment at a new, higher level.

The complexity of climate change impacts on delta environments, and their severity for human populations are well exemplified by the Nile Delta. Alexandria, Rosetta, and Port Said are at particular risk, and even a sea level rise of 50 centimetres could mean that 2 million people would have to abandon their homes. Even without accelerating sea level rise, the Nile Delta has suffered erosion through the building of dams which has prevented river sediment reaching the delta. Under former, natural conditions, Nile sediments, on reaching the sea, generated sand bars and dunes which contributed to delta accretion. About a century ago, an inverse process was initiated and the delta began to retreat. For example, the Rosetta mouth lost about 1.6 kilometres of its length from 1898 to 1954. The imbalance between sedimentation and erosion appears to have started with the delta barrages (1861) and then been continued by later works, culminating with the Aswan High Dam. In addition, large amounts of sediment are retained in an extremely dense network of irrigation and drainage channels that has been developed in the Nile Delta itself. Much of the Egyptian coast is now 'undernourished' with sediment and, as a result of this overall erosion of the shoreline, the sand bars bordering Lake Manzala and Lake Burullus on the seaward side are becoming eroded and likely to collapse. If this were to happen, the lakes would be converted into marine bays, so that saline water would come into direct contact with low-lying cultivated land and freshwater aquifers.

Coral reefs are thought to be one of the most notable coastal geomorphological hotspots in the face of future climate change, because of their sensitivity to sea level, sea surface temperature, ocean acidity, and storms. Ocean acidification is currently regarded as a major threat to coral reefs over the coming century. A proportion of the extra carbon dioxide being released into the atmosphere by the burning of fossil fuels and biomass is absorbed by sea water. As this combines with water, it produces carbonic acid. An increase in carbonic acid in sea water will cause it to become more acidic. Several centuries from now, if we continue to add carbon dioxide to the atmosphere, oceans will be more acidic than at any time in the past 300 million years. This will be harmful to those organisms like corals that depend on the presence of carbonate ions to build their skeletons out of calcium bicarbonate. As it is, many coral reefs have also been damaged by human activities over the past century and many are now in a highly vulnerable condition. Furthermore, episodic natural events such as hurricanes, and in particular El Niño events, have had a major effect on many coral reef landscapes. Whilst coral reefs are undoubtedly threatened by climate change, it is possible to overstate the threat and underestimate the potential for reefs to adapt to new conditions.

Desert areas, especially desert margins, are also thought to be one of the major geomorphological hotspots in the face of climate change. The major impact of climate change will be on sediment mobilization, especially in terms of wind erosion. Changes in climate could affect wind erosion either through their impact on erosivity or through their effect on erodibility (or both). Erosivity refers to the ability of winds to cause erosion, and is controlled by a range of wind variables including velocity, frequency, duration, and turbulence. Unfortunately, general circulation models as yet give little indication of how wind characteristics might be modified in a warmer world, so that prediction of future changes in wind erosivity is problematic. Erodibility refers to the sensitivity of a surface to erosion, and is largely controlled by vegetation cover and surface material characteristics, both of which can be influenced markedly by climate. In general, vegetation cover, which protects the ground surface and modifies the wind regime, decreases as conditions become more arid. Likewise, climate affects surface materials by controlling their moisture content, the nature and amount of clay mineral content (cohesiveness), and organic levels.

If soil moisture levels decline as a result of changes in precipitation and/or temperature, there is the possibility that dust storm activity could increase in a warmer world. A comparison between the Dust Bowl years of the 1930s and model predictions of precipitation and temperature for the Great Plains of Kansas and Nebraska indicates that mean conditions could be similar to or even worse than those of the 1930s under enhanced greenhouse conditions. If dust storm activity were to increase as a response to global warming, it is possible that this could have a feedback effect on precipitation that would lead to further decreases in soil moisture. However, the impact and occurrence of dust storms will depend a great deal on land-management practices, and recent decreases in dust storm activity in North Dakota and in parts of the High Plains have resulted from conservation measures.

Sand dunes, because of the crucial relationships between vegetation cover and sand movement, are highly susceptible to the effects of climate change. Some areas, such as the south-west Kalahari or portions of the High Plains of the USA, may be especially prone to the effects of changes in precipitation and/or wind velocity because of their location in climatic zones that are close to a climatic threshold between dune stability and activity. The development in the use of optical dating of sand grains and studies of explorers' accounts has led to the realization that such marginal dune fields have undergone repeated phases of change at decadal and century timescales in response to extended drought events during the course of the Holocene. This reminds us that the past is a vital archive of information for us in predicting the future for many landscapes.

Perhaps the most detailed scenarios for dune remobilization by global warming have been developed for the Kalahari in southern Africa. Much of the mega-Kalahari is currently vegetated and stable, but general circulation models suggest that by 2099 all dune fields, from South Africa in the south to Zambia and Angola in the north, will be reactivated because of a diminished vegetation cover and drier sediments. The consequences of dune encroachment and reactivation could be serious and might lead to a loss of agricultural land, the overwhelming of buildings, roads, canals, runways, and the like, abrasion of structures and equipment, damage to crops, and the impoverishment of soil structure.

Permafrost areas are also seen as important landscape hotspots under climate change. Permafrost is defined as ground which remains frozen for at least two consecutive years. Usually, an 'active layer' which thaws seasonally is found above true permafrost. In extremely cold areas, the permafrost is called 'continuous' (because it underlies more than 90% of the surface), whereas towards the warmer margins it is called 'discontinuous' (occupying between 50% and 90% of the land) or 'sporadic' (underlying less than 50% of the surface). Discontinuous and sporadic types are most prone to climate change impacts, as much of the permafrost is within a couple of degrees of thawing. Permafrost covers some 25% of the Earth's land surface, mainly in the northern hemisphere in areas such as northern Russia, Canada, Alaska, and Greenland. For example, permafrost covers 85% of the land surface of Alaska and 50% of Canada. Many indigenous peoples live in permafrost areas, and in northern Russia there are many large cities and ports underlain by permafrost. Already, warming has been rapid within these areas, and predictions are for further amplification of temperatures here. There is historical evidence that permafrost can degrade speedily when warmed. For instance, during the warm 'climatic optimum' of the Holocene (c. 6,000 years ago), the southern limit of discontinuous permafrost in the Russian Arctic was up to 600 kilometres north of its present position. Similarly, researchers have demonstrated that along the Mackenzie Highway in Canada, between 1962 and 1988, the southern fringe of the discontinuous zone had moved north by about 120 kilometres in response to an increase over the same period of 1°C mean annual temperature. Strong future warming in high latitudes will have a similarly marked impact on permafrost. In Canada, over half of what is now the discontinuous zone could be eliminated, so that the boundary between continuous and discontinuous permafrost might shift northwards by hundreds of kilometres.

Degradation of permafrost may lead to an increasing scale and frequency of slope failures, particularly in mountainous areas.

Thawing reduces the strength of both ice-rich sediments and frozen jointed bedrock. Water released by melting may further promote mass movements, while increases in the thickness of the zone subject to thaw – the active layer – may make more material available for debris flows. Coastal bluffs may be subject to increased rates of erosion if they suffer from thermal erosion caused by permafrost decay. This would be accelerated still further if sea ice were to be less prevalent, for sea ice can protect coasts from wave erosion and debris removal. Local coastal losses to erosion of as much as 40 metres per year have been observed in some locations in both Siberia and Canada in recent years, while erosive losses of up to 600 metres over the past few decades have occurred in Alaska.

One of the severest consequences of permafrost degradation is ground subsidence and the formation of closed depressions called 'thermokarst'. This is likely to be a particular problem for engineering structures in the zones of relatively warm permafrost (the discontinuous and sporadic zones), and where the permafrost is rich in ice. With thawing, ice-rich areas will settle more than those with lesser ice contents, producing irregular hummocks and depressions. Water released from ice melt may accumulate in such depressions. The depressions may then enlarge into thaw lakes as a result of thermal erosion. Once they have started, thaw lakes can continue to enlarge for decades to centuries because of wave action and continued thermal erosion of the banks. However, if thawing eventually penetrates the permafrost, drainage occurs that leads to ponds drying up.

From our consideration of coastal, desert, and permafrost landscape hotspots, it is clear that climate change is likely to have major impacts. The confidence with which we can predict what will happen relates to the amount of information we have about the past history of some of these sensitive landscapes, often gained through applying some of the range of techniques we introduced in Chapter 3. We also need to have very good knowledge of how the various, and often highly complex, landscapes function today. It is only through rigorous application of geomorphological science that we can improve our predictions of how landscapes will change. However, this is only one half of the story. Whilst the impacts of climate change on landscapes are evident from the examples presented above, it is important to consider whether there are any geomorphic feedbacks which will link landscape change back to climate change. What we have already seen is that many impacts of climate change on landscape affect, often in quite complex ways, both vegetation and geomorphology. In some cases, these impacts set in train feedback loops which then have knock-on effects on climate.

Two important feedbacks involve dust and methane. Looking first at dust, it is clear that climate change or human pressures on the land surface may cause wind erosion of surface materials in drylands to increase. An intriguing question is the extent to which this may have feedback effects on global climate. There are various ways in which this might occur. First, the amount of dust in the atmosphere affects the radiation budget of the Earth, either by absorbing or scattering incoming radiation from the Sun. This could cause some cooling to occur. Second, dust contains nutrients, such as iron and phosphorus, so that if these are deposited over the oceans they will serve to fertilize ocean biota, especially phytoplankton. An increase in the mass of these biota would draw down carbon dioxide from the atmosphere. It might also stimulate production of dimethyl sulphide (DMS), which might increase cloud albedo (reflectivity). Both of these processes could lead to cooling. In addition, the amount of dust in the atmosphere may affect the size and number of cloud condensation nuclei, and so may affect rainfall levels. Some scientists fear that dust suppresses rainfall and so may accentuate drought.

Feedbacks involving methane are also important. The melting of permafrost as a result of global warming may have a series of positive feedback effects that may cause further global warming to

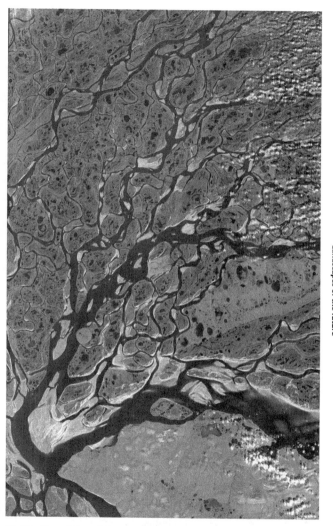

14. The Lena River delta, Russia, from Google Earth, illustrating the high number of thaw lakes

occur. First, the permafrost is a frozen reservoir of preserved roots and leaves. When the permafrost melts, it is like turning off a refrigerator crammed full of food. Putrefaction occurs that releases the carbon stored in the organic matter. Second, the melting of permafrost creates huge numbers of thaw lakes (thermokarst), and methane is generated in these lakes and thence enters the atmosphere. Methane is a highly effective greenhouse gas. Third, methane is stored in the Arctic Ocean floor as a frozen ice-and-methane mixture. If the ocean warms up and the submarine permafrost starts to melt, this mixture will become unstable and release prodigious amounts of methane into the atmosphere. The Lena River delta, shown in Figure 14, is a good example of an environment in which melting of thaw lakes is occurring, potentially contributing to global warming.

We now know a substantial amount about how landscapes respond to climate change, but there is still much more to be discovered. Whilst governments and inter-governmental organizations urgently require answers to some key questions, geomorphologists and other scientists find it increasingly hard to provide concrete and simple answers. There are many reasons for this, notably because landscapes are complex and respond to climate changes in inherently complicated ways. Similarly, feedbacks within the linked ocean-atmosphere-terrestrial system (some of which we have discussed above) make predictions incredibly difficult. Furthermore, climate is not the only influence on landscapes. Over the next few decades and centuries, local human impacts on landscapes will be hugely important, not only those that happen in the future, but also those that have already occurred and are conditioning how the landscapes behave. We must not forget tectonic activity as well, which will undoubtedly influence many landscapes over the coming years, perhaps working in tandem with climatic change and local human impacts.

Chapter 8
Landscapes, art, and culture

Humans have a visceral connection with their natural surroundings, often reflected in a love of landscape. Feeling for landscape can simply result in active enjoyment, but can also be translated into art, poetry, and other cultural affirmations of scenery and landforms, as well as into tangible interventions such as carvings, buildings, and structures. Such appreciation of landscapes is also formalized in attempts to conserve sites for future generations. We explore a range of these responses to landscape in this chapter, moving from active enjoyment to artistic and cultural expressions, and, finally, to landscape conservation. The term 'cultural geomorphology' is a useful one to encapsulate the scientific study of these varied aspects of human engagement with the landscape, and we begin this chapter with a brief consideration of what cultural geomorphology is all about.

The term 'cultural geomorphology' has recently been proposed by Ken Gregory to encompass cultural reactions to and perceptions of landscape, and how these should be considered by geomorphologists, especially in terms of improving environmental management. The key issue is that human societies around the world interact with their natural environment in a range of different ways, highly influenced by cultural perspectives. If we want to contemplate manipulating these engagements with landscape in order to manage, improve, or conserve the environment, then it is crucial to incorporate some understanding and appreciation of the cultural dimension. For example, in projects which aim to improve the management of river systems, it is important to understand the cultural relationship that people in the area have had with their rivers. This allows us to disentangle why they have pursued particular management solutions and what types of management might work in the future. Where river waters are of religious significance, such as in parts of India, river-management strategies need to take this into account. Such cultural dimensions are also vital to successful conservation of landscapes.

A vast number of people enjoy dramatic landscapes and scenery and respond to it in an active rather than necessarily creative way. From the 18th-century Grand Tour to the 21st-century Gap Year, people have travelled widely and have visited many key landscape sites around the world, as well as nearer to home. Many millions of people around the world enjoy spectacular landscapes, as exemplified by the millions who visit the national parks of the USA each year (Table 3). These parks are essentially landform spectacles rather than merely biosphere reserves. The same applies to some of the great national parks of Australia – Uluru (Ayers Rock), Geikie Gorge, Windjana Gorge. In some countries, a dramatic landform may be one of the prime attractions for tourists. This is the case with the Victoria Falls in Zimbabwe and Zambia, Table Mountain in South Africa, the Sugar Loaf in Rio, and Wadi Rum in Jordan.

Many outdoor pursuits are based upon landforms – such as caving and potholing, white-water rafting, bog snorkelling, and mountain climbing. Others, such as orienteering and geocaching, enable many to explore and start to understand topography and scenery. The most experienced and well-travelled cavers and mountaineers also reflect on the beauty and majesty of the scenery within which they explore. Joe Simpson, the mountaineer who nearly died on

Table 3. Numbers of visitors to ten major national parks in the USA in 2008 (in millions) (Data from National Park Service)

Great Smoky Mountain, South Appalachians	9.0
Grand Canyon (Arizona)	4.4
Yosemite (California)	3.4
Olympic (Washington)	3.1
Yellowstone (Wyoming, Montana, and Idaho)	3.1
Cuyahoga Valley (Ohio)	2.8
Rocky Mountain (Colorado)	2.8
Zion (Utah)	2.7
Grand Teton (Wyoming)	2.5
Acadia (Maine)	2.1

Siula Grande in the Andes, writes about the Himalayas as seen from the Hunza valley as 'Pristine, untouchable. Lofting into the sky, perfectly beautiful' in *Touching the Void* (1988).

Geomorphologists themselves are far from immune from enjoying the beauty of the landscapes in which they work. John Wesley Powell, for example, wrote thus of the impact upon his senses of the Grand Canyon:

The Grand Canyon of the Colorado is a canyon composed of many canyons. It is a composite of thousands of gorges. In like manner, each wall of the canyon is a composite structure, a wall composed of many walls, but never a repetition. Every one of these almost innumerable gorges is a world of beauty in itself. In the Grand Canyon there are thousands of gorges like that below Niagara Falls, and there are a thousand Yosemites. Yet all the canyons unite to form one Grand Canyon, the most sublime spectacle on the earth.

J. W. Powell, *Canyons of the Colorado* (New York: Flood and Vincent, 1895), p. 390

In like vein, the soldier and explorer R. A. Bagnold became the greatest aeolian geomorphologist of the 20th century because of the spell that 'promiscuous barchans' and other dunes of the Libyan Desert cast upon him:

Instead of finding chaos and disorder, the observer never fails to be amazed at a simplicity of form, an exactitude of repetition, and a genetic order unknown in nature on a scale larger than that of crystalline structure. In places vast accumulations of sand weighing millions of tons move inexorably, in regular formation over the surface of the country, growing, retaining their shape, even breeding, in a manner which by its grotesque imitation of life is vaguely disturbing to an imaginative mind.

R. A. Bagnold, *The Physics of Blown Sand and Desert Dunes* (London: Methuen, 1941), p. xxi

Some geomorphologists have gone even further. Vaughan Cornish, a British geographer who did geomorphological work on waves as well as a very wide range of other geographical studies in the early 20th century, also wrote books entitled *The Poetic Expression of Natural Scenery* in 1931, and *The Beauties of Scenery* in 1943.

Throughout the history of humanity, such engagements with landscape have resulted in artistic outputs. The genre of landscape painting contains many different examples of representational images of landscapes, from the early Chinese dynasties through to the 18th- and 19th-century Western landscape art of Thomas Gainsborough, J. M. W. Turner, and others. In China, the tower karst landscapes (often called 'stone forests' by Chinese scientists and scholars) have been depicted in many paintings, especially since the late Tang dynasty, around AD 700 onwards. Sometimes, art and science have gone hand in hand, as seen in the work of Leonardo da Vinci, who produced many landscape drawings and sketches. In the *Codex Leicester* (1508–12), he outlines his global theory, based on his observations, in which he suggested that the Earth was a cavern filled with water, or as he puts it:

> The great elevations of the peaks of mountains above the sphere of the water may have resulted from the fact that a very large portion of the earth which was filled with water, that is to say, the vast cavern inside the earth, may have fallen in a vast part of its vault towards the centre of the earth, being pierced by means of the course of the springs which continually wear away the place where they pass ... It is of necessity that there should be more water than land, and the visible portion of the sea does not show this; so that there must be a great deal of water inside the earth, besides that which rises into the lower air and which flows through the rivers and springs.

(*Codex Leicester*, <http://www.universalleonardo.org/trail.php?trail=198&work=325>)

More recently, landscape has come to be actively used in art in a more physical and non-representational way, such as in the land art movement. This began in the late 1960s and is characterized by works such as Robert Smithson's 'spiral jetty' constructed as an earthwork on the Great Salt Lake, Utah, in 1970, from around 5,900 metric tons of soil and basalt from the site. Once created, such earthworks become a part of the landscape, subjected to natural processes and influencing them in turn. Salts have crystallized around the spiral since it was constructed and are now part of the artwork. An even more dramatic earthwork is Michael Heizer's 'double negative' created in 1969 at Mormon Mesa in Nevada. This required the dynamiting and bulldozing of 218,000 metric tons of rock and debris to create two deep trenches, together 457 metres long, on either side of a retreating mass movement on the edge of a sandstone escarpment. Such giant artworks are reminiscent of ancient creations such as the Nazca lines, or the more recent carvings at Mount Rushmore. More modest and ephemeral, but no less intriguing, land art pieces include those of Richard Long, who creates sculptures such as 'a circle in the Andes' (1972) from stones, boulders, or boot prints as part of walks (which he also records and documents). These perhaps echo the many examples of rock art – paintings, carvings, and constructions – found throughout the world dating from prehistoric and from more recent times. Non-Western cultures have also developed artistic forms involving natural landforms, as depicted in Figure 15.

Many landscapes have more permanent and shared cultural significance in the form of religious sites. Sometimes the physical landscape itself is of religious significance, as, for example, in Uluru in Australia, Mount Fuji in Japan, Mount Taishan, Shandong Province, China, and other sacred mountains. In other instances, it is the ecological component of the landscape that is of significance, such as the sacred groves of India. Alternatively, the landscape may be the site where religious buildings and structures are built, or where significant paintings or carvings are to be found. For example, many caves are the sites of important religious images and relicts, whilst many monasteries and other religious buildings are located in remarkable and often remote landscapes which themselves are often of religious significance. St Catherine's monastery is located at the foot of Mount Sinai in the Sinai desert, Egypt, where according to the Bible Moses received the Ten Commandments from God (Figure 16). A complex of religious

15. A historic Chinese 'geosculpture' in the Forbidden City, Beijing

buildings (gates, pavilions, archways, and kiosks) is found on Mount Taishan in China, which has been the site of Taoist and earlier styles of worship for over three millennia and has also provided important inspiration for Chinese landscape painters over many centuries.

16. St Catherine's monastery, Mount Sinai, Egypt

Landscapes can also be significant in terms of defence, wars, and the establishment of political systems. The lineaments of Hadrian's Wall, for example, are provided by an intrusion of igneous rocks, the Whin Sill, which is a classic example of an igneous landform. Castles, hill forts, burial mounds, and other structures are also often found in distinctive landscape contexts, such as on the edge of escarpments, on the tops of cliffs and hills, or other areas that provided good visibility over long distances for the original inhabitants. Looking at a broader scale, the nature of terrain sets the scene for many economic activities and indeed settlement and political systems. Worldwide, many villages and larger settlements, often now abandoned, have also been constructed in intimate relation with the natural environment, such as the cave dwellings in consolidated volcanic ash in Cappadocia, Turkey, the temples and buildings carved out of sandstone at Petra in Jordan, and the buildings of Chaco Canyon, New Mexico, USA. Many of these important items of cultural heritage owe their location, form, and purpose to the landscape which surrounds them, and the geomorphological canvas on which they are written is often a key part of their value today.

The presence of culturally important landscapes has led to a worldwide effort to conserve them in the face of development pressures. Landscape conservation has a long history locally, nationally, and globally. Local conservation efforts have often focused on conserving some key species, or area of natural beauty, or perhaps a landform, archaeological site, or individual historic building, rather than looking at the landscape in a more holistic way. In the UK, there are now, for example, a network of RIGS (regionally important geological/geomorphological sites). Early local protection efforts often occurred because of the input of interested national historians, such as the protection of the 'Pierre-à-Bot' glacial erratic boulder in Neuchâtel, Switzerland, in 1838. This was recommended for protection by the local authorities by Louis Agassiz, who had used it as an illustration of the former extent of the Rhone Glacier in his work. The world's first national park was established in 1872 at Yellowstone, USA, when 2.2 million acres of wilderness were set aside '... for the enjoyment of the people'. In Britain, national parks were not formally designated until the 1950s, following a long campaign by many individuals and groups, including the geographer Vaughan Cornish. National parks are usually set in landscapes of great physical natural beauty, with additional interesting wildlife and cultural heritage, although the balance between these three elements varies considerably. The UK now has 15 national parks, China has 187, and the USA nearly 400. Globally, there are now over 113,000 protected areas of this type, covering 149 million square kilometres, or 6% of the Earth's land surface (according to the UK National Parks website).

International efforts to conserve landscape are most clearly seen in the UNESCO World Heritage List. This is an initiative aimed at the

> identification, protection and preservation of areas, places, and objects around the globe which are considered of outstanding universal value for humankind, and which have to be passed on to future generations according to the 'Convention concerning the Protection of the World Cultural and Natural Heritage'.

> (UNESCO, 1972)

There are three types of world heritage site: cultural, natural, and mixed. As of 2009, the list contains 890 sites, of which 689 are cultural, 176 natural, and 25 mixed. In order to become listed as a natural world heritage site, one or more of the following criteria must be satisfied:

- to contain superlative natural phenomena or areas of exceptional natural beauty and aesthetic importance (criterion no. 7);
- to be outstanding examples representing major stages of Earth's history, including the record of life, significant ongoing geological processes in the development of landforms, or significant geomorphic or physiographic features (criterion no. 8);
- to be outstanding examples representing significant ongoing ecological and biological processes in the evolution and development of terrestrial, fresh water, coastal, and marine ecosystems and communities of plants and animals (criterion no. 9);
- to contain the most important and significant natural habitats for in-situ conservation of biological diversity, including those containing threatened species of outstanding universal value from the point of view of science or conservation (criterion no. 10).

To give some illustrations: the Iguazu/Iguaçu Falls on the boundary of Argentina and Brazil have been listed on criterion no. 7, whilst landscapes such as the 'Jurassic coast' in East Devon and Dorset, UK, and Yosemite, USA, have been listed on grounds of criterion no. 8. Many other sites listed on criteria nos. 9 and 10 also have extraordinary physical (geomorphological) characteristics as well as their important ecological nature. Mixed sites often have hugely important natural landscapes, such as the Göreme National Park and the Rock Sites of Cappadocia in Turkey, and Mount Huangshan, China, as do many cultural sites, such as Petra in Jordan, and Tadrart Acacus in Libya. Indeed, in the case of Petra (rock-hewn buildings) and Tadrart Acacus (rock art), the cultural heritage would not have been created without the availability of the natural materials and environment. As the geomorphologist Piotr Migoń puts it, 'many World Heritage sites are geomorphological icons, long used as the best possible illustrations of surface processes at work and insights into the geomorphic history of the earth's surface'.

Whilst landscape is clearly an important component of the above conservation efforts, there have been many moves within recent years to focus effort more explicitly on geoconservation (that is, the conservation of geological and geomorphological heritage, or 'geoheritage'). The reasoning behind these moves is that whilst biodiversity conservation and the conservation of cultural heritage have become hugely successful, the other dimensions of the landscape have become neglected. As, in most cases, cultural, biological, and geomorphological/geological heritage are highly intertwined, this separation does not make much sense. More than this, there are some important aspects of geoheritage which demand conservation in their own right – such as abandoned quarries that cut through and exhibit important phases of recent Earth history, for example. Unless geoconservation is seen as an important goal in its own right, there is a danger that such sites will become marginalized or destroyed. Aligned to ideas of geoconservation has been the concept of 'geodiversity', which, in many ways, parallels that of biodiversity. Murray Gray has defined geodiversity as: 'The natural range (diversity) of geological (rocks, minerals, fossils), geomorphological (land form, processes) and soil features. It includes their assemblages, relationships, properties and systems.' There have, however, been many arguments about how closely linked biodiversity and geodiversity actually are. One of the problems is that both of them are complex concepts, especially geodiversity, and thus quite hard to pin down, measure, and compare. How can we make the case that geoconservation and geodiversity are important?

The beauty of landscape is something that appeals to many people worldwide, whether or not they have an appreciation of the geomorphological science which underpins this beauty. Art and culture continue to be heavily influenced by landscape, but such artistic aspects remain largely divorced from the science of geomorphology. More seriously, the underlying aspects of the landscape (geology and geomorphology) are often sadly underestimated by conservation bodies, even though these are often critical to shaping and characterizing the entire landscape, including its cultural heritage. Geomorphologists, perhaps working within the guise of cultural geomorphology, can undoubtedly help boost the profile of geoconservation and work with ecologists and heritage scientists to provide a more balanced approach to identifying and conserving landscapes of outstanding value at local, national, and global levels.

Chapter 9
Unseen landscapes

Many landscapes remain unseen and unexplored until new techniques become available which enable us to visit and visualize them. Over the past few hundred years, our knowledge of landscapes on Earth has been revolutionized with the advent of new modes of transport such as ocean-going vessels, aeroplanes, four-wheel-drive vehicles, submarines, and satellites. Little of the land surface is now really unseen, although cave systems continue to be discovered and explored. The last great unexplored and largely unseen realm on Earth is that of the ocean floor, which is now being revealed to us through improved techniques such as ship-borne multibeam sonar bathymetry and 3-D seismic survey. The landscapes of other planets also remain largely unseen until space exploration vehicles visit them and send back information from a range of sensors. In this chapter, we introduce the recent research that is now allowing us to see some of these exciting landscapes, focusing on the submarine landscapes of Earth, and the planetary landscapes of Mars and Titan. This research illustrates how we interpret new landscapes through our knowledge of landscapes closer to home, and, in turn, how our observations of new landscapes help us refine our interpretations of more familiar landscapes.

The ocean floor remains largely unexplored directly by humans, apart from some submarine-based expeditions, but its major outlines have now been mapped using a range of remote-sensing techniques, both ship-based and using satellites. The ship-based techniques utilize sound waves to sense the ocean floor, whilst satellites collect altimetry data using microwave radar. Sonar was first used during the First World War. In essence, echosounding works by sending a pulse of sound out and timing its return after it has been reflected by the sea floor in order to calculate the depth of water. Over time, sonar techniques have improved, and now they can reveal much about not only the depth but also the characteristics (such as sediment composition) of the sea floor. Ship-based echosounding has been used to create bathymetric data for many well-visited parts of the oceans over the past 30 or so years, but this is time-consuming because of the relatively slow speed at which ships travel. Satellite altimetry has been used to fill in the gaps. It uses microwave radar to measure accurately the height of the sea surface, from which the sea floor topography can be calculated. Simply put, the ocean surface reflects the Earth's gravity field and, in turn, the Earth's gravity field is affected by undersea topography. For example, a large undersea mountain affects the gravity field, causing a local 'pimple' in the sea surface. The variations in sea surface topography are tiny, but can be measured using microwave radar altimetry. Together, ship-based and satellite techniques have been used to create global topographic maps (or DEMs) of the ocean floor at a horizontal resolution of around 2 kilometres. Together with terrestrial topographic data (derived from Shuttle Radar Interferometry, or SRTM), this has been used to produce the ETOPO1 world topographic map at kilometre-scale resolution. Such datasets have been included in Google Earth, for example.

Whilst the basic lineaments of the ocean floor have been identified using the techniques described above, concurrent developments in technology have permitted detailed, high-resolution views of the sea floor. Two particular types of technique are worth introducing here. The first technique, swath (or swathe) bathymetry, has helped produce high-resolution digital elevation models of the sea floor. Swath bathymetry builds on the echosounding technique and uses an array of sound pulses sent out in a fan shape below and to the sides of a vessel. Two types of technique can be used to create a swath, multibeam or interferometric sonar. Typical multibeam arrays use around 190 beams and can map a swath of several kilometres wide. The second set of techniques allows the creation of 'images' of the characteristics of surface and subsurface materials using sidescan sonar and 3-D seismic survey. Sidescan sonar collects information on the intensity of sound received by echosounding from a submarine vehicle towed behind a ship. The intensity of sound (backscatter) relates to the type of surface, with, for example, high backscatter (strong reflections) produced by boulders and weaker reflections from muddy sediments. Seismic survey goes one step further and investigates the subsurface interactions, with an acoustic signal producing a cross-section through the sea floor. Different types of sediment will interact with the sound wave in different ways (such as reflection and refraction), producing recognizable patterns. Surveys are run by ships along individual tracks, producing 2-D images. Where a series of tracks are run in close proximity, 3-D images can be produced. Much of the highest-resolution bathymetric and seismic data is collected for oil exploration and production purposes, and is often not easily available because of commercial sensitivities.

Techniques that allow visualization of the shape of the sea floor and the nature of the materials which underlie it are highly useful for geomorphologists and marine geologists who want to explore the landforms, processes, and history of the ocean floor. The term 'sea-floor geomorphology' is sometimes used to describe such research which has both practical (oil exploration, hazard assessment) and academic aspects. As predicted by plate tectonics theory, the ocean floor is geologically young (mostly less than 150 million years) compared with the continents. A vast range of interesting relief features are found on the sea floor, reflecting the impact of tectonic, glacial, mass movement, and other processes. At the large scale, for example, we can discriminate mid-ocean ridges, oceanic trenchs, continental shelf, and seamounts, amongst other landforms.

Let's start by looking at seamounts. Whilst Charles Darwin wrote about seamounts and their importance to marine biodiversity, our knowledge of their nature and number has been quite poor until very recently. Seamounts are submarine volcanoes, which rise at least 1,000 metres above the surrounding sea floor. Work by John Hillier and Tony Watts from Oxford University using ship-track bathymetry has recently identified over 200,000 submarine volcanoes – an order of magnitude larger than previous estimates. Other, smaller-scale, landforms recently revealed by high-resolution bathymetry and seismic survey include a host of glacial features within fjords. For example, the tidewater glaciers in Svalbard have produced a range of submarine glacial landforms revealed by 3-D seismic survey, such as mega-scale glacial lineations, which are elongated ridges of glacial sediments. The complex, over-printed nature of the glacial landforms on the floor of fjords can be used to infer the nature and rate of glacial surging activity over the past few hundred years.

One of the most exciting areas of subsea geomorphological research in recent years has been the identification and interpretation of subsea mass movements. Landslides occur on the ocean floor in similar ways to those that occur on land, with creep, flow, slump, slide, and fall types all common. One type of mass movement unique to the subsea environment is the turbidity current, which involves the rapid movement downslope of sediment-rich water which is more dense and turbid than the surrounding water. Turbidity currents are often triggered by earthquakes and other subsea mass movements. Subsea mass movements are often immense. One of the largest documented mass movements is the Agulhas submarine landslide off South Africa which is reported to have involved some 20,000 cubic kilometres of sediment and took place across a run-out distance of more than 140 kilometres. Several factors can trigger subsea mass movements, and can be grouped into those that reduce the strength of the body of material involved and those that increase the stresses upon it. Important triggers can be earthquakes, hurricanes, storm waves, over-steepening, gas hydrate disassociation, or glacial loading. As on land, subsea mass movements are often complex events, involving a number of linked movements of different types, often producing long-term instability and dynamism. The Storegga Slide (see box) is one of the largest and best-studied subsea landslide complexes on Earth.

The Storegga Slide

Storegga is Old Norse for the 'Great Edge', and the great edge referred to is the edge of the continental shelf off Norway. Here, because of steep underwater terrain, susceptible glacial marine sediments, occasional earthquakes, and release of submarine methane, there are conditions that promote instability. Some major submarine landslips have occurred here, with perhaps more than 20 failures during the last 2.6 million years. The most famous slide, which is thought to date back to just over 8,000 years ago, occurred about 120 kilometres off the Norwegian coast. It was on a truly enormous scale. An area the size of Iceland slumped down, involving an estimated 290–320 kilometre length of coastal shelf. The run-out distance across the ocean floor was around 800 kilometres. This is almost the length of mainland Britain. Not only is the morphology and scale of this event of great interest, but it also has more general significance. First, given the frequency of past events, it is likely that such events may occur in the future as well. Second, the site of the slides is also that of the great Ormen Lange offshore gas field. Third, the Storegga Slide created an enormous tidal wave, or tsunami, the impact of which was felt as far away as the British Isles. In the Shetland Islands, the tsunami reached onshore heights at least 20 metres above the sea level of that time.

The risk of tsunamis resulting from subsea landslides has been a major trigger for research on them within recent years. Landslides in fjords and on the flanks of oceanic islands are especially likely to affect humans. Islands chains such as the Canary Islands possess many subsea landslides. Bathymetric survey has helped to identify a large landslide on the sea floor adjacent to Tenerife, for example. The landslide covers an area of around 5,500 square kilometres (much larger than the island itself), is around 100 kilometres in length, and involves an estimated 1,000 cubic kilometres of material. The form of the landslide indicates that it has, in fact, been created in a series of events. The nature of the steeply sloping Canary Islands means that they are thought to be prone to such events, and some scientists have suggested that a future large event could create a huge tsunami capable of devastating the Caribbean and eastern USA seaboard. However, this is a worst-case scenario, and the evidence from Tenerife indicates that multiple small failures are more likely than one massive event.

Other previously unseen worlds whose landscapes are becoming rapidly visible thanks to new imagery are the other planets and moons within our Solar System. Scientists have, for centuries, speculated about the nature of planetary surfaces based on the views provided by the naked eye and telescopes. Today, planetary geomorphology is a strong and vibrant area of study in its own right, enhanced by rapidly developing observations from spacecraft and contributing to human exploration of other planets and their resources. The nature of planetary landscapes is controlled by their location relative to the Sun, conditions of gravity and atmosphere, and the materials that make up the planetary surface, which all affect the balance of processes operating. Earth, for example, is affected by tectonic, volcanic, and denudational processes and, to a much less obvious degree, impact cratering. Some of these processes are limited on other planetary bodies. For example, there is no concrete evidence of plate tectonics on other planets, and the extremely thin atmosphere on some planetary bodies, such as the Moon, restricts the activity of many denudational processes. Impact cratering is a very obvious component of many planetary landscapes, especially where atmospheres are thin, as on the Moon.

Mars has had a long fascination for scientists, and has many similarities in terms of atmosphere and surface landscapes to Earth, despite being further from the Sun, smaller, and possessing much lower gravity. A selection of key comparisons between the present conditions on Earth and those on Mars is presented in Table 4.

Table 4. Comparisons between Earth and Mars

	Earth	Mars
Distance from Sun (million km)	150	228
Diameter (km)	12756	687
Day length	24 hours	24 hours 40 mins (= 1 sol)
Year length	365 days	669 sols
Gravity (ms^{-2})	9.81	0.38
Atmospheric pressure (mb)	1000	6
Mean surface temperature (°C)	15	−23
Atmospheric composition	78% N_2, 21% O_2, 0.035% CO_2, H_2O 0–4%	95.32% CO_2, N_2 2.7%, O_2 0.13%, H_2O 0.03%

The first images of Mars came from the Mariner missions in the 1960s which flew past, but it wasn't until 1976 that the Viking missions provided detailed information from orbiting and landing on the planet. More than 50,000 photos were sent back from the Viking missions, illustrating the diversity of landscapes on Mars. A number of NASA missions in the 1990s and 2000s (Mars Global Surveyor, Mars Pathfinder, Mars Odyssey, Mars Explorer, and Mars Reconnaissance Orbiter) have added vastly to the information on the planet, with imagery, spectrometric data, laser altimetry, and other data produced. The European Space Agency has also had a successful mission in the form of Mars Express, which has been orbiting Mars since 2004, with high-resolution stereo camera (HRSC) and mineralogical mapping spectrometer (OMEGA) equipment on board. The resolution and quality of the data produced in recent years are very impressive, often being comparable with those produced by satellites orbiting Earth. The Mars Reconnaissance Orbiter, which was launched in 2005, produces imagery at resolutions of around 1 metre from the HIRISE camera and spectrometry data from the CRISM instrument at unprecedented resolution with pixels of 10s to 100s of metres.

As well as information from satellites, there have been successful lander vehicle missions on Mars, starting with the Pathfinder lander in 1997. More recently, Spirit and Opportunity rover vehicles landed in 2004 and, at the time of writing, have been active and collecting data for over 2,000 sols (Martian days). Phoenix landed in 2008 and sent back data from the northern polar area for five months before becoming frozen. The vehicles that land on the surface and move around have provided remarkable data on their surroundings from onboard cameras (such as high-resolution stereo imagery from PANCAM). They have also used tools such as the rock-abrasion tool, or RAT, and the microscopic imager, or MI, to view the surface and subsurface of boulders and rock outcrops to a resolution of about 100 microns. For every successful mission to Mars, there

Dust on Mars

The 'Red Planet' could just as easily be called the 'Dust Planet', for yellowish-brown dust gives Mars its distinctive colour. Dust storms occur almost daily, with thousands occurring each year. Telescopic observations since the 18th century and images delivered by spacecraft missions have shown Mars to be an arid planet dominated by the presence of dust both suspended in the atmosphere and deposited widely over the planet's surface. Features like linear streaks, up to 400 kilometres in length, are indicators of the power of dust entrainment.

Dust events on Mars have been observed at all scales, ranging from local dust devils to storms that envelop the entire planet, dubbed 'global dust storms'. These planet-encircling events occur approximately one year in three, usually in the late southern spring when Mars is closest to the Sun. During the Martian summer, in the lower boundary layer of its clear, thin, cold atmosphere, the large temperature gradient that exists above the relatively warm surface may support intense free convection and the formation of dust devils, which are often greater than those found on Earth, reaching several hundred metres across and 8 kilometres high. Regional dust storms may be produced whenever the poleward temperature gradient is sufficiently large to generate intense zonal circulation across the mid-latitudes in the form of baroclinic waves. Other regional dust storms are produced by winds descending from high relief. Regional dust storms affect the radiation budget and this can lead to feedback effects that may cause the development of dust storms of global dimensions. Such an event was detected by the Hubble space telescope in June 2001. What was to become the biggest event for about a quarter of a century began as a small dust cloud inside the Hellas Basin (a deep impact crater in Mars's southern hemisphere). By early July, the dust cloud had spilled out of the basin and engulfed the whole planet. It is possible that airborne dust particles absorb sunlight and warm the Martian atmosphere strongly in their vicinity. Warm pockets of air spread quickly towards colder regions, thereby generating strong winds. These lift more dust off the ground and so create a positive feedback. In this model, dust heating seems to play a broadly analogous role to the release of latent heat in moist convection during the development of tropical storms and hurricanes on Earth.

have been many failures, including the loss of the Beagle 2 lander which crash-landed from the Mars Express mission.

So what do the landscapes of Mars look like? There are two major types of terrain: the heavily cratered uplands found around the equator and southern hemisphere; and the northern lowland plains whose surface has been influenced by ice, volcanic, and aeolian processes. It is clear that, like Earth, Mars has experienced major shifts in climate over long timespans, whilst, unlike Earth, there is no convincing evidence that Mars has a plate tectonic system. Much of the Martian surface is covered by basaltic lavas. Mars contains some of the largest volcanoes within the Solar System, including Olympic Mons, which at 624 kilometres in diameter and 25 kilometres high is many times the size of the largest volcano on Earth, Mauna Loa in Hawaii, which is 10 kilometres high and 120 kilometres in diameter. There are two large polar ice caps on Mars, as well as extensive subsurface permafrost. Several large canyons are found on Mars, notably Valles Marineris, which is around 4,000 kilometres long and up to 7 kilometres deep, and which appears to have been formed by crustal swelling and collapse. Many kinds of mass movement are observed on Mars, many of them extremely large – such as the impressive features found in Valles Marineris. Here, landslides have developed on the side of the canyon walls, with fall heights of around 6.5 kilometres and sediment volumes of around 10^{24} cubic kilometres – orders of magnitude higher than terrestrial landslides, but comparable with some of the largest subsea features on Earth. Other impressive landforms on Mars are dunes, including a suite of linear and barchan features which can be ten times the size of similar features on Earth. Indeed, many weird and wonderful dune types have been identified on Mars imagery (see Figure 17), prompting comparative studies using Google Earth and other remotely sensed imagery from Earth to look for similar features. Because of the low atmospheric pressures and generally low wind speeds on Mars, dunes are not very active, especially in

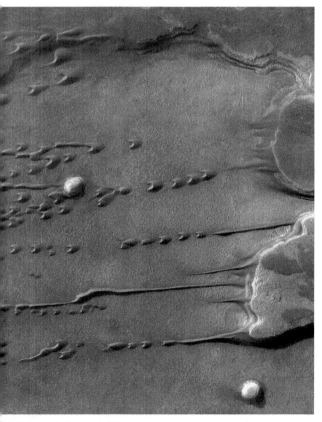

17. Enigmatic dunes on Mars

polar areas, where there are also seasonal frosts of water and carbon dioxide which help cement the dunes. Landforms on Mars are large relative to those on Earth because of the much lower acceleration due to gravity on Mars.

The landforms on Mars that have excited most interest are those that are hypothesized to have been produced by fluids such as water or carbon dioxide. Gullies, for example, are found on many crater walls and may result from movements of sediment mixed with water or other fluids. Channels are also visible on Mars. One of the largest channel systems on Mars is Nirgal Vallis. There has been much debate amongst geomorphologists about how such channel networks were formed – by the action of surface running water, groundwater sapping, or other processes? Spectrometry data (from, for example, OMEGA and CRISM) have confirmed the presence of water ice on Mars, as well as the presence of minerals that form under aqueous conditions (such as carbonates). It is now known that there are extensive subsurface ice deposits on Mars, as well as the polar ice caps, and that the planet has experienced major climatic swings in the past, causing repeated ice ages. So, some of these features could have been formed in the distant past by water, or could be formed by certain other fluids more recently.

Whilst imagery and other data from Mars are an invaluable source of information, answering questions about the Martian landscape has relied heavily on the use of studies of similar (often called 'analogue') landscapes on Earth, as well as modelling and experimentation. For example, studies of channel formation on Mars have relied on observations of gullies and valleys created by groundwater sapping in Egypt and on the Colorado plateau in the USA. It has proved difficult, however, to confirm the origin of such features on Earth, let alone extrapolate to Mars. Recent studies involving cosmogenic isotope dating and detailed field observations at Box Canyon, Idaho, however, make a plausible case for catastrophic flooding as the major agent carving out this canyon in basalt, rather than groundwater sapping. Their findings may have great relevance for Mars.

Geomorphologists have recently become excited by the landscape on a body much further from the Sun. Titan is the largest of the moons of Saturn and has recently become visible geomorphologically, as a result of the ongoing Cassini mission, which first visited Titan in 2004, dropped the Huygens probe onto the surface in 2005, and has flown by several times since then. Titan is comparable in size to our Moon, but is atmospherically more like Earth. Table 5 illustrates some basic comparisons between Titan and the Earth.

Table 5. Comparisons between Earth and Titan

	Earth	Titan
Distance from Sun (million km)	150	1,427
Diameter (km)	12756	5150
Day length	24 hours	16 days
Year length	365 days	29.5 years
Gravity (ms^{-2})	9.81	0.137
Atmospheric pressure (mb)	1000	1500
Mean surface temperature (°C)	15	−180
Atmospheric composition	78% N_2, 21% O_2, 0.035% CO_2, H_2O 0–4%	98.4% N_2, 1.4% CH_4, traces of hydrocarbons

The images of the surface of Titan sent back by the Cassini mission, including those from the Huygens probe, illustrate a landscape quite similar to that on Earth – except the surface is made up of water ice, not rock, and sculpted by liquid methane, not water. The very cold temperatures, relative to those on Earth, facilitate the occurrence of liquid methane and water ice. Rain is thought to occur as giant flakes of methane, falling through the low-gravity, dense atmosphere. River channels, lakes, and dunes are all easily visible from radar imagery (see Figure 18 for an example). Lakes are common in the polar regions, whilst large expanses of linear dunes cover about 20% of Titan's surface around the equator. The dunes appear to be composed of sand-sized organic particles and are of a scale similar to large linear dunes on Earth in places

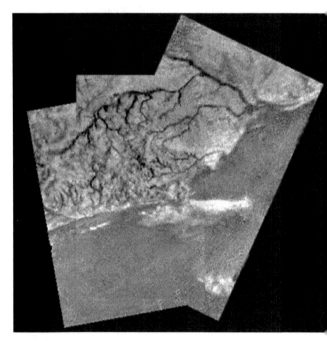

18. **Radar images of channels on Titan**

such as Namibia. The Huygens probe landed within what appeared to be a floodplain, with rounded cobbles composed of water ice visible nearby indicating fluvial transport. Radar imagery from its descent shows a channel network very similar to those on Earth, albeit carved by liquid methane out of ice.

Our glimpses of new geomorphologies on the sea floor, Mars, and Titan have been facilitated by rapidly developing observational techniques. In many ways, the landforms we can now see in such remote areas are very similar to those on the terrestrial surface of Earth, but also show some tantalizing differences. These new observations have reinvigorated geomorphology, giving it some new questions to tackle and also throwing fresh light on some more familiar features on Earth. The exploration of subsea geomorphology also has many practical applications, in terms of enhancing exploitation of offshore resources such as oil and mineral nodules and reducing the threat of hazards such as subsea mass movements. Similarly, the observations of landscapes on Mars and Titan help with questions of the habitability of other planets.

Further reading

A. D. Abrahams and A. J. Parsons, *Geomorphology of Desert Environments*, 2nd edn. (Dordrecht: Springer, 2009). A comprehensive edited volume on desert geomorphology.

D. Anderson, A. S. Goudie, and A. G. Parker, *Global Environments through the Quaternary* (Oxford: Oxford University Press, 2007). A survey of the major environmental changes of the last 3 million years.

R. A. Bagnold, *The Physics of Blown Sand and Desert Dunes* (London: Methuen, 1941). The classic work on sand dunes.

R. P. Beckinsale and R. J. Chorley, *The History of the Study of Landforms or the Development of Geomorphology*: Volume 3, *Historical and Regional Geomorphology 1890–1950* (New York: Routledge, 1991). Part of a magisterial four-volume series on the history of geomorphology.

D. I. Benn and D. J. A. Evans, *Glaciers and Glaciation* (London: Arnold, 1998). A very full analysis of glacial landforms.

E. C. F. Bird, *Coastal Geomorphology: An Introduction* (Chichester: Wiley, 2000). A worldwide introduction to the geomorphology of coastlines.

W. Bland and D. Rolls, *Weathering: An Introduction to the Scientific Principles* (London: Arnold, 1998). An introduction to the processes that cause rocks to decay and disintegrate.

J. S. Bridge, *Rivers and Floodplains: Forms, Processes, and Sedimentary Record* (Oxford: Blackwell Science, 2003). A modern treatment of many aspects of fluvial geomorphology.

D. Brunsden, R. Gardner, A. S. Goudie, and D. Jones, *Landshapes* (Newton Abbot: David and Charles, 1988). An accessible account of geomorphology.

D. W. Burbank and R. S. Anderson, *Tectonic Geomorphology: A Frontier in Earth Science* (Malden, Mass.: Blackwell Science, 2001). An innovative discussion of how landforms and tectonics are related.

T. P. Burt et al. (eds.), *The History of the Study of Landforms*, Volume 4 (London: The Geological Society, 2008). A history of geomorphology between the 1890s and the 1960s.

D. R. Butler, *Zoogeomorphology: Animals as Geomorphic Agent* (Cambridge: Cambridge University Press, 1995). The fundamental text on how animals operate as geomorphological agents.

R. Charlton, *Fundamentals of Fluvial Geomorphology* (London: Routledge, 2008). A simple introductory text on the geomorphology of rivers.

R. J. Chorley, R. P. Beckinsale, and A. J. Dunn, *The History of the Study of Landforms or the Development of Geomorphology*: Volume 2, *The Life and Work of William Morris Davis* (London: Methuen, 1973). Part of a major survey of the history of geomorphology that concentrates on W. M. Davis.

R. J. Chorley, A. J. Dunn, and R. P. Beckinsale, *The History of the Study of Landforms*, Volume 1 (London: Methuen, 1964). The first instalment of a history of geomorphology, looking at the early days.

R. U. Cooke and J. C. Doornkamp, *Geomorphology in Environmental Management*, 2nd edn. (Oxford: Oxford University Press, 1990). A pioneering discussion of how geomorphologists can contribute to environmental management.

R. U. Cooke, A. Warren, and A. S. Goudie, *Desert Geomorphology* (London: UCL Press, 1993). A comprehensive overview of the geomorphology of deserts.

M. J. Crozier, *Landslides: Causes, Consequences and Environment* (London: Routledge, 1989). A fundamental treatment of mass movements.

R. Dikau, D. Brunsden, L. Schrott, and M.-L. Ibsen, *Landslide Recognition* (Chichester: Wiley, 1996). A major survey of the diversity of mass movement types.

P. G. Fookes, E. M. Lee, and G. Milligan (eds.), *Geomorphology for Engineers* (Dunbeath: Whittles Publishing, 2005). A multi-author survey of how geomorphology can contribute to civil engineering.

D. C. Ford and P. W. Williams, *Karst Geomorphology and Hydrology*, 2nd edn. (Chichester: Wiley, 2007). The best modern text on limestone geomorphology.

P. Francis and C. Oppenheimer, *Volcanoes* (Oxford: Oxford University Press, 2004). A well-illustrated analysis of volcanic landforms.

H. M. French, *The Periglacial Environment*, 2nd edn. (Harlow, Essex: Addison Wesley Longman, 1996). A very useful survey of landforms in periglacial areas.

H. M. French (ed.), *Periglacial Geomorphology*, (Geomorphology: Critical Concepts in Geography, Volume V) (London: Routledge, 2004). A collection of classic papers.

D. Gillieson, *Caves: Processes, Development and Management* (Oxford: Blackwell, 1996). An introduction to caves.

A. S. Goudie, *The Landforms of England and Wales* (Oxford: Blackwell, 1990). A survey of the geomorphology of part of the British Isles.

A. S. Goudie (ed.), *Geomorphological Techniques*, 2nd edn. (London and New York: Routledge, 1994). A multi-author handbook to the techniques used by geomorphologists.

A. S. Goudie, *Great Warm Deserts of the World: Landscapes and Evolution* (Oxford: Oxford University Press, 2002). A regional treatment of deserts and their landscapes.

A. S. Goudie (ed.), *Encyclopedia of Geomorphology* (London: Routledge, 2004). A two-volume guide to modern geomorphology.

A. S. Goudie, *The Human Impact*, 6th edn. (Oxford: Blackwell, 2006). An introduction to how humans have transformed their environment.

A. S. Goudie and N. J. Middleton, *Desert Dust in the Global System* (Heidelberg: Springer, 2006). A discussion of dust storms and their importance.

A. S. Goudie and H. A. Viles, *Salt Weathering Hazards* (Chichester: Wiley, 1997). An example of a geomorphological hazard that has particular significance for the built environment.

M. Gray, *Geodiversity: Valuing and Conserving Abiotic Nature* (Chichester: Wiley, 2003). A broad introduction to geodiversity, the threats to it, and how to manage them.

M. Gutiérrez, *Climatic Geomorphology* (Amsterdam: Elsevier, 2005). A survey by a distinguished Spanish geomorphologist describing how different climatic areas have different landforms.

R. J. Huggett, *Fundamentals of Geomorphology*, 2nd edn. (Abingdon: Routledge, 2007). A very useful introduction.

Further reading

I. Livingstone and A. Warren, *Aeolian Geomorphology: An Introduction* (Harlow, Essex: Longman, 1996). A basic introduction to the role of wind as a geomorphological agent.

G. Masselink and M. G. Hughes, *An Introduction to Coastal Processes and Geomorphology* (London: Arnold, 2003). An introduction to the coast.

P. Migoń, *Granite Landscapes of the World* (Oxford: Oxford University Press, 2007). A discussion of how granites have distinctive landscapes.

R. P. C. Morgan, *Soil Erosion and Conservation*, 3rd edn. (Oxford: Blackwell, 2005). The standard work on soil erosion.

D. J. Nash and S. J. McLaren (eds.), *Geochemical Sediments and Landscapes* (Oxford: Blackwell, 2007). An edited collection of chapters on weathering crusts and related features.

P. N. Owens and O. Slaymaker (eds.), *Mountain Geomorphology* (London: Hodder Headline, 2004). An edited collection on the geomorphological character of mountains.

O. Slaymaker, T. Spencer, and C. Embleton-Hamann (eds.), *Geomorphology and Environmental Change* (Cambridge: Cambridge University Press, 2009). A major statement on how global change affects the geomorphic environment.

M. A. Summerfield, *Global Geomorphology: An Introduction to the Study of Landforms* (Harlow, Essex: Longman, 1991). An excellent survey of geomorphology, especially at the global scale.

M. F. Thomas, *Geomorphology in the Tropics: A Study of Weathering and Denudation in Low Latitudes* (Chichester: Wiley, 1994). The best treatment of the geomorphology of the humid tropics.

K. J. Tinkler, *A Short History of Geomorphology* (London: Croom Helm, 1985). A neat summary of the history of geomorphology.

C. R. Twidale and J. R. Vidal Romani, *Landforms and Geology of Granite Terrains* (Leiden: Balkema, 2005). A very full discussion of the distinctive nature of granite landforms.

H. A. Viles (ed.), *Biogeomorphology* (Oxford: Blackwell, 1988). The first survey of how geomorphology and the biosphere are related.

H. A. Viles and T. Spencer, *Coastal Problems* (London: Arnold, 1995). An analysis of the contribution that geomorphologists make to understanding coastal problems.

H. J. Walker and W. E. Grabau, *The Evolution of Geomorphology: A Nation-by-Nation Summary of Development* (Chichester: Wiley, 1993). This work shows the diversity of approaches to geomorphology.

C. D. Woodroffe, *Coasts: Form, Process and Evolution* (Cambridge: Cambridge University Press, 2002). An excellent and comprehensive survey of coastal geomorphology.

R. W. Young, A. L. Wray, and A. R. M. Young, *Sandstone Landforms* (Cambridge: Cambridge University Press, 2009). The best analysis of what makes the geomorphology of sandstone areas very distinctive.

Index

CLASSICS
A Very Short Introduction
Mary Beard and John Henderson

This Very Short Introduction to Classics links a haunting temple on a lonely mountainside to the glory of ancient Greece and the grandeur of Rome, and to Classics within modern culture – from Jefferson and Byron to Asterix and Ben-Hur.

'The authors show us that Classics is a "modern" and sexy subject. They succeed brilliantly in this regard … nobody could fail to be informed and entertained – and the accent of the book is provocative and stimulating.'

John Godwin, *Times Literary Supplement*

'Statues and slavery, temples and tragedies, museum, marbles, and mythology – this provocative guide to the Classics demystifies its varied subject-matter while seducing the reader with the obvious enthusiasm and pleasure which mark its writing.'

Edith Hall

MUSIC
A Very Short Introduction
Nicholas Cook

This stimulating Very Short Introduction to music invites us to really *think* about music and the values and qualities we ascribe to it.

'A *tour de force*. Nicholas Cook is without doubt one of the most probing and creative thinkers about music we have today.'

Jim Samson, University of Bristol

'Nicholas Cook offers a perspective that is clearly influenced by recent writing in a host of disciplines related to music. It may well prove a landmark in the appreciation of the topic … In short, I can hardly imagine it being done better.'

Roger Parker, University of Cambridge

www.oup.co.uk/vsi/music

PSYCHOLOGY
A Very Short Introduction
Gillian Butler and Freda McManus

Psychology: A Very Short Introduction provides an up-to-date overview of the main areas of psychology, translating complex psychological matters, such as perception, into readable topics so as to make psychology accessible for newcomers to the subject. The authors use everyday examples as well as research findings to foster curiosity about how and why the mind works in the way it does, and why we behave in the ways we do. This book explains why knowing about psychology is important and relevant to the modern world.

'a very readable, stimulating, and well-written introduction to psychology which combines factual information with a welcome honesty about the current limits of knowledge. It brings alive the fascination and appeal of psychology, its significance and implications, and its inherent challenges.'

Anthony Clare

'This excellent text provides a succinct account of how modern psychologists approach the study of the mind and human behaviour. ... the best available introduction to the subject.'

Anthony Storr

www.oup.co.uk/vsi/psychology

POLITICS
A Very Short Introduction
Kenneth Minogue

In this provocative but balanced essay, Kenneth Minogue discusses the development of politics from the ancient world to the twentieth century. He prompts us to consider why political systems evolve, how politics offers both power and order in our society, whether democracy is always a good thing, and what future politics may have in the twenty-first century.

'This is a fascinating book which sketches, in a very short space, one view of the nature of politics ... the reader is challenged, provoked and stimulated by Minogue's trenchant views.'

Ian Davies, *Talking Politics*

'a dazzling but unpretentious display of great scholarship and humane reflection'

Neil O'Sullivan, University of Hull

ARCHAEOLOGY
A Very Short Introduction
Paul Bahn

This entertaining Very Short Introduction reflects the enduring popularity of archaeology – a subject which appeals as a pastime, career, and academic discipline, encompasses the whole globe, and surveys 2.5 million years. From deserts to jungles, from deep caves to mountain tops, from pebble tools to satellite photographs, from excavation to abstract theory, archaeology interacts with nearly every other discipline in its attempts to reconstruct the past.

'very lively indeed and remarkably perceptive ... a quite brilliant and level-headed look at the curious world of archaeology'

Barry Cunliffe, University of Oxford

'It is often said that well-written books are rare in archaeology, but this is a model of good writing for a general audience. The book is full of jokes, but its serious message – that archaeology can be a rich and fascinating subject – it gets across with more panache than any other book I know.'

Simon Denison, editor of *British Archaeology*

www.oup.co.uk/vsi/archaeology

ONLINE
CATALOGUE
A Very Short Introduction

Our online catalogue is designed to make it easy to find your ideal Very Short Introduction. View the entire collection by subject area, watch author videos, read sample chapters, and download reading guides.

http://fds.oup.com/www.oup.co.uk/general/vsi/index.html

SOCIAL MEDIA
Very Short Introduction

Join our community
www.oup.com/vsi

- Join us online at the official Very Short Introductions **Facebook** page.
- Access the thoughts and musings of our authors with our online **blog**.
- Sign up for our monthly **e-newsletter** to receive information on all new titles publishing that month.
- Browse the full range of Very Short Introductions online.
- Read **extracts** from the Introductions for free.
- Visit our library of **Reading Guides**. These guides, written by our expert authors will help you to question again, why you think what you think.
- If you are a teacher or lecturer you can order inspection copies quickly and simply via our website.